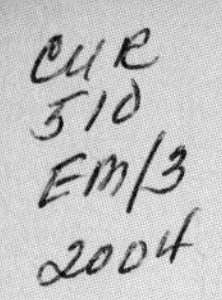

Everyday Mathematics®

Student Math Journal 1

The University of Chicago
School Mathematics Project

Wright Group

The McGraw-Hill Companies

UCSMP Elementary Materials Component

Max Bell, Director

Authors

Max Bell
Jean Bell
John Bretzlauf*
Amy Dillard*
Robert Hartfield
Andy Isaacs*
James McBride, Director
Kathleen Pitvorec*
Peter Saecker

Technical Art

Diana Barrie*

**Second Edition only*

Photo Credits

Phil Martin/Photography, Jack Demuth/Photography, Cover Credits: Sand, starfish, orange wedges, crystal/Bill Burlingham Photography, Photo Collage: Herman Adler Design Group

Contributors

Carol Arkin, Robert Balfanz, Sharlean Brooks, Ellen Dairyko, James Flanders, David Garcia, Rita Gronbach, Deborah Arron Leslie, Curtis Lieneck, Diana Marino, Mary Moley, William D. Pattison, William Salvato, Jean Marie Sweigart, Leeann Wille

Send all inquiries to:
Wright Group/McGraw-Hill
P.O. Box 812960
Chicago, IL 60681

Printed in the United States of America.

ISBN 0-07-584483-4

8 9 10 11 12 DBH 10 09 08 07 06 05

Contents

Unit 1: Routines, Review, and Assessments

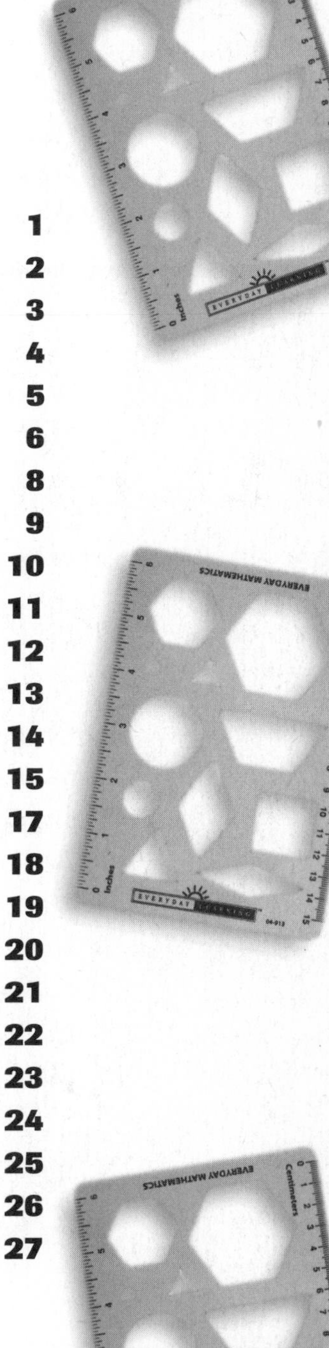

A note at the bottom of each journal page indicates when that page is first used.
Some pages will be used again during the course of the year.

Unit 2: Adding and Subtracting Whole Numbers

Unit 3: Linear Measures and Area

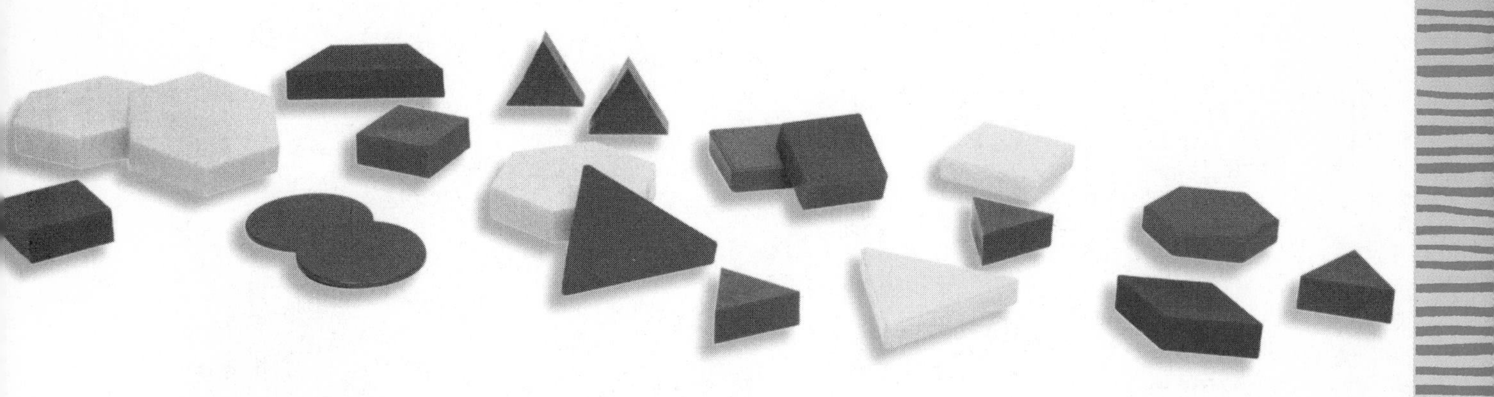

Unit 4: Multiplication and Division

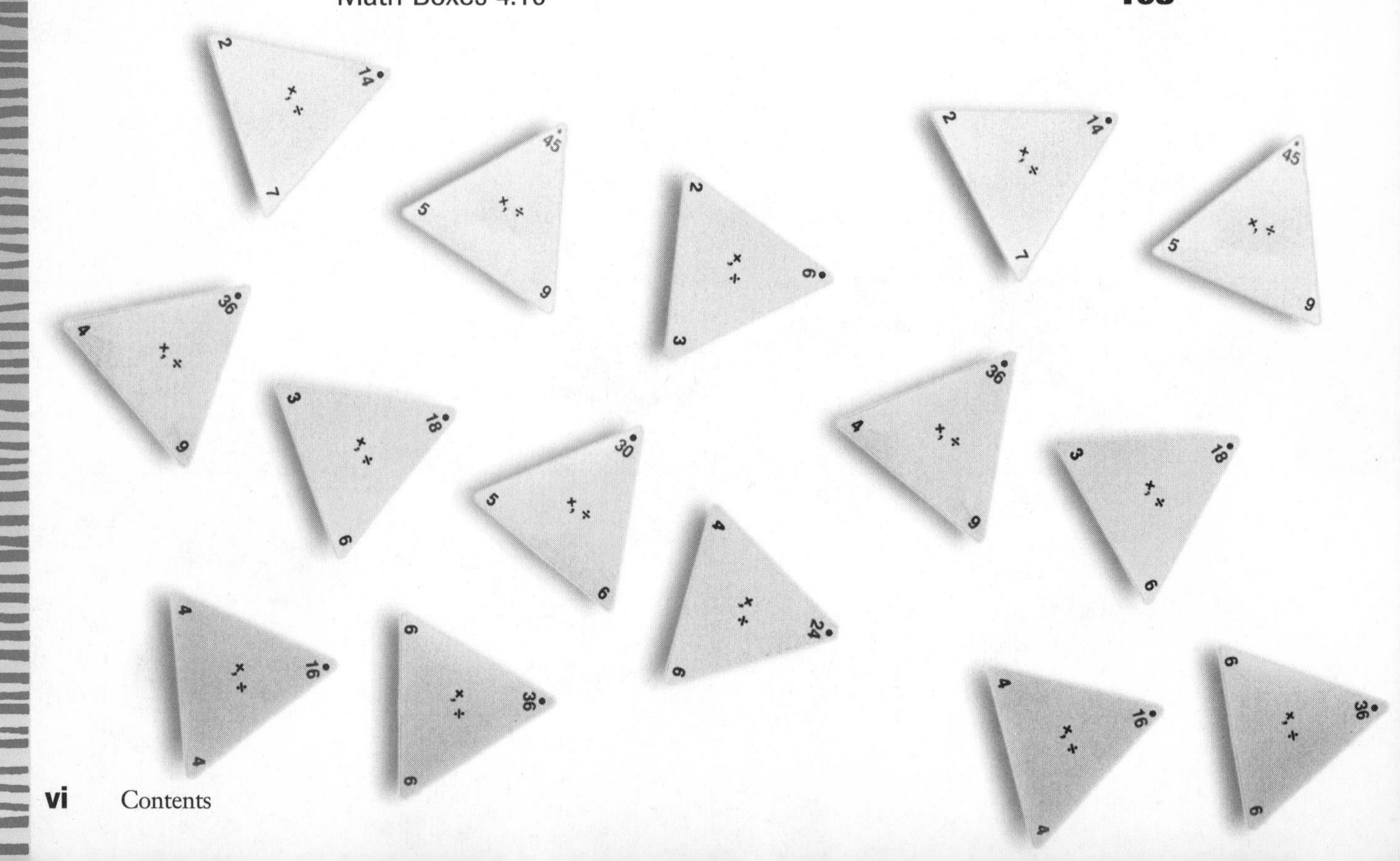

Unit 5: Place Value in Whole Numbers and Decimals

Unit 6: Geometry

Activity Sheets

Date Time

A Numbers Hunt

Look for numbers in your classroom. Write the numbers in the table. Look for numbers that you cannot "see" but you can find by counting or measuring. Record these numbers, too.

Number	**Unit** (if there is one)	**What does the number mean?**	**How did you find the number?** (count, measure, another way?)
Example: 16	*Crayons*	*Tells how many crayons are in a box*	*Number is on the box*
Example: 30	*Inches*	*Height of my desk*	*Measured my desk*

Date Time

Number-Grid Puzzles

1. Complete the grid.

541			544						550
551		553			556			559	
	562			565					570
			574			577			
581				585			588		
		593						599	
	602				606				
			614						620

Fill in the missing numbers.

2.

	69
78	

3.

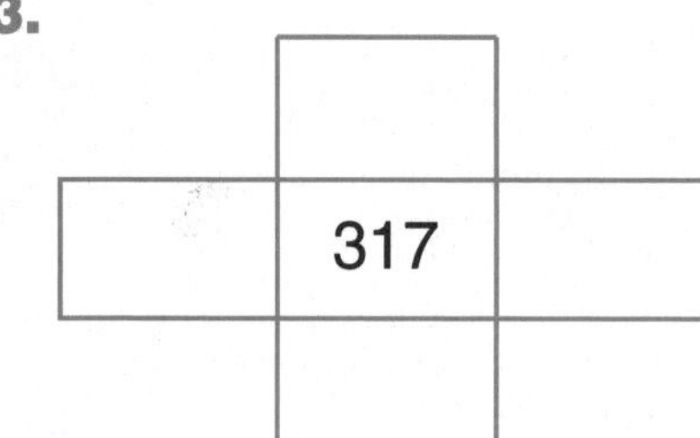

4.

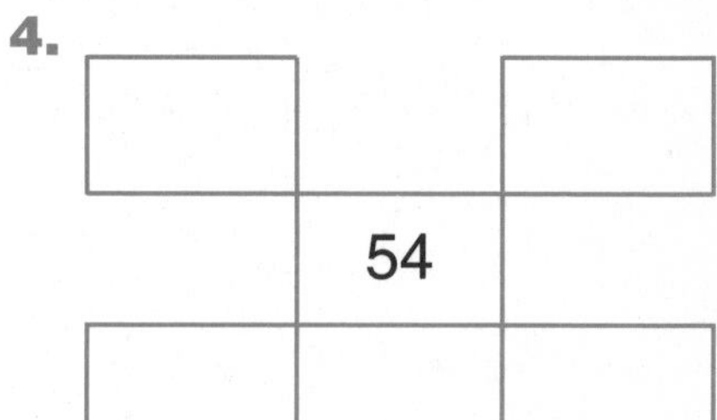

5.

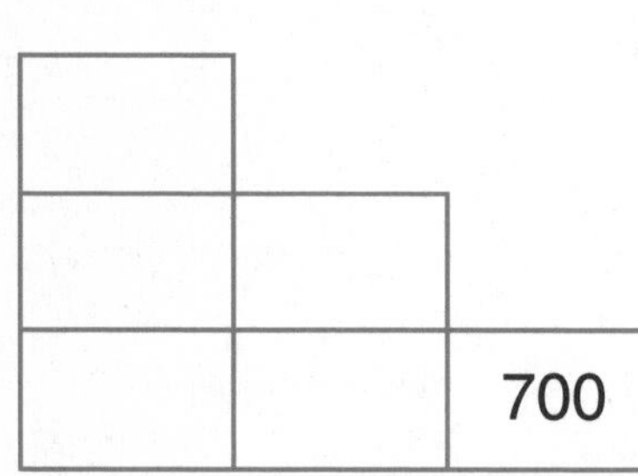

6.

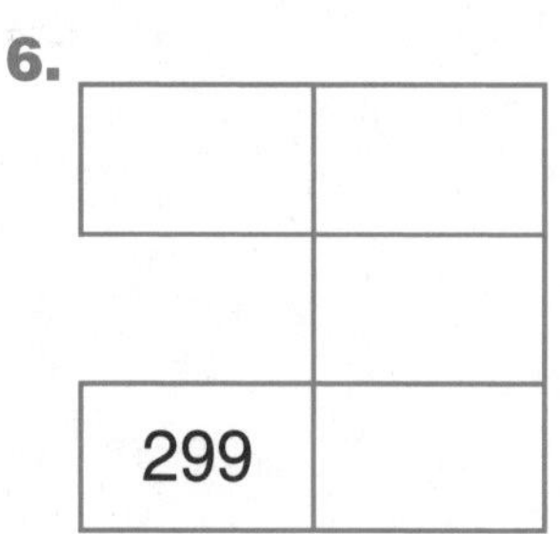

7.

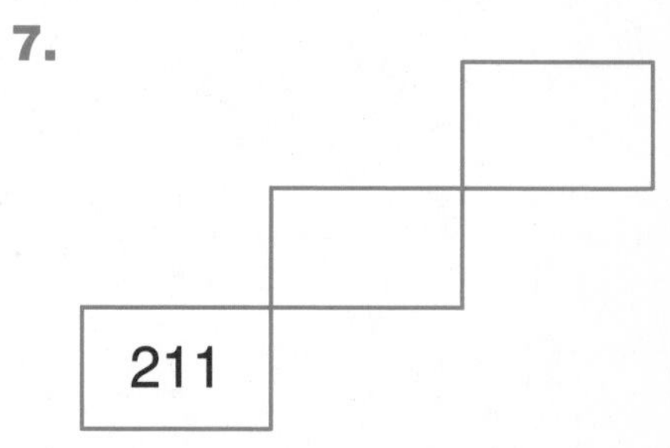

Make up your own puzzles. Ask someone to solve them.

8.

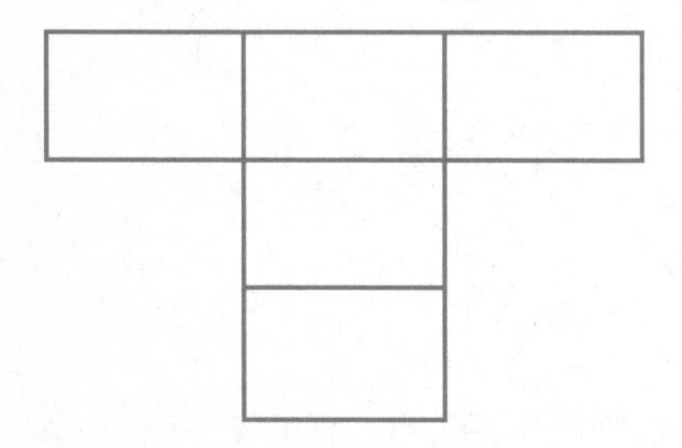

9.

Use with Lesson 1.2.

Date Time

Looking up Information

Math Message

1. Turn to page 270 in your *Student Reference Book.*

   How many yards are there in 1 mile? ________ yards

Work with a partner. Use your *Student Reference Book* for Questions 3–6.

2. Write your partner's first name. ____________________

   Write your partner's last name. ____________________

3. Look up the word **circumference** in the Glossary. Copy the definition.

   ____

4. Read the essay "Tally Charts."

   a. Then solve the Check Your Understanding problems.

      Problem 1: ____________________

      Problem 2: ____________________

   b. Check your answers in the Answer Key.

   c. Describe what you did to find the essay.

      ____

      ____

5. Find the Measurement section. Which of the following units of length is about the same length as a person's height?

   a. yard b. thumb c. fathom d. cubit

   On which page did you find the answer? ________

6. Look up the rules of the game *Less Than You!* P your partner.

Using Mathematical Tools

In Problems 1 and 2, record the time shown on the clocks. In Problem 3, draw the minute hand and the hour hand to show the time.

1.

____

2.

____

3.

6:10

Use your ruler.

4. Measure the line segment. about ________ inches

5. Draw a line segment 10 centimeters long.

vour calculator to do these problems.

573 + 859 + 6,051 = ________

7. 20,748 − 8,967 = ________

38 = ________

9. 1,978 ÷ 23 = ________

ern-Block Template to draw the following shapes:

11. a hexagon

12. a trapezoid

blems 10–12 are quadrangles?

Date ______________ Time ______________

Math Boxes 1.5

1. What is today's date?

____

What will be the date in 6 days?

____

What will be the date in 1 week?

____

2. Fill in the missing numbers.

	174		
		205	

3. Write the number that is 10 more.

42 ________

160 ________

901 ________

Write the number that is 10 less.

59 ________

120 ________

4. Count back by 3s.

42, ____, ____, 33,

____, ____, ____, ____,

____, ____, ____, ____,

____, ____

5. About what time is it?

____

6. Add.

9 + 0 = ________

1 + 7 = ________

____

____

7 +

Displaying Data

1. How many first names are there? __________

2. How many last names are there? __________

3. With which names will you work—first names or last names? ________________

4. Make a tally chart for your set of names.

____________ Names	
Number of Letters	**Number of Children**
2	
3	
4	
5	
6	
7	
8	
9	
10 or more	

any letters does the longest name have? ________________ letters
ber of letters in the longest name is called the **maximum.**

etters does the shortest name have? ________________ letters
f letters in the shortest name is called the **minimum.**

of the numbers of letters? ________________ letters
member what the range is,
ent Reference Book.)

of data? ________________ letters

Date Time

Displaying Data (cont.)

9. Make a bar graph for your set of data.

Title: ______________________

Date Time

Name-Collection Boxes

1. Write 10 names in the 20-box.

20

2. Write 10 names in the 24-box.

24

3. Three names do not belong in this box. Cross them out. Then write the name of the box on the tag.

fourteen

10 + 6

10 less than 26

10 − 6

8 twos

4 + 4 + 4

half of 32

10 + 2 − 4 + 6 − 8 + 10

4. Make up your own box.

Date Time

Math Boxes 1.6

1. Complete the pattern.

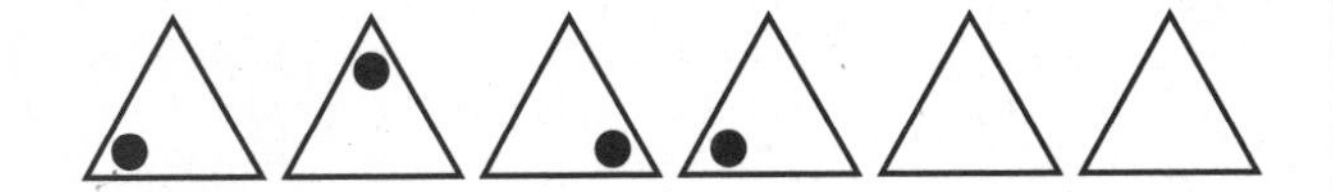

2. 6,347

What value does the 6 have? ________

What value does the 7 have? ________

What value does the 3 have? ________

What value does the 4 have? ________

SRB 18 19

3. Use Ⓟ, Ⓝ, Ⓓ, and Ⓠ.

Show $0.89 in two ways.

4. How many trees have exactly 6 bugs? ________

How many trees have exactly 3 bugs? ________

Number of Bugs per Tree	Number of Trees
2	//
3	/
4	////
5	/
6	//

5. Count by 10s.

23, ______, ______, 53,

______, ______, ______, ______,

______, ______, ______, ______,

______

6. Add.

$4 + 8 =$ ________

________ $= 9 + 2$

$4 + 3 =$ ________

$5 + 5 =$ ________

________ $= 8 + 8$

Date Time

Finding Differences

									0
1	2	3	4	5	6	7	8	9	10
11	12	13	14	15	16	17	18	19	20
21	22	23	24	25	26	27	28	29	30
31	32	33	34	35	36	37	38	39	40
41	42	43	44	45	46	47	48	49	50
51	52	53	54	55	56	57	58	59	60
61	62	63	64	65	66	67	68	69	70
71	72	73	74	75	76	77	78	79	80
81	82	83	84	85	86	87	88	89	90
91	92	93	94	95	96	97	98	99	100
101	102	103	104	105	106	107	108	109	110

Use the number grid to help you solve these problems.

1. Which is less, 83 or 43? ________ How much less? ________

2. Which is less, 33 or 78? ________ How much less? ________

3. Which is more, 90 or 55? ________ How much more? ________

4. Which is more, 44 or 52? ________ How much more? ________

Find the **difference** between each pair of numbers.

5. 71 and 92 ________

6. 26 and 46 ________

7. 30 and 62 ________

8. 48 and 84 ________

9. 43 and 60 ________

10. 88 and 110 ________

Date Time

Skip Counting on the Number Grid

1. Start at 0 and count by 4s on the number grid.
Mark an X through each number in your count.

2. Start at 0 again and count by 5s on the number grid.
Draw a circle around each number in your count.

									0
1	2	3	4	5	6	7	8	9	10
11	12	13	14	15	16	17	18	19	20
21	22	23	24	25	26	27	28	29	30
31	32	33	34	35	36	37	38	39	40
41	42	43	44	45	46	47	48	49	50
51	52	53	54	55	56	57	58	59	60
61	62	63	64	65	66	67	68	69	70
71	72	73	74	75	76	77	78	79	80
81	82	83	84	85	86	87	88	89	90
91	92	93	94	95	96	97	98	99	100
101	102	103	104	105	106	107	108	109	110
111	112	113	114	115	116	117	118	119	120
121	122	123	124	125	126	127	128	129	130
131	132	133	134	135	136	137	138	139	140
141	142	143	144	145	146	147	148	149	150
151	152	153	154	155	156	157	158	159	160
161	162	163	164	165	166	167	168	169	170
171	172	173	174	175	176	177	178	179	180
181	182	183	184	185	186	187	188	189	190
191	192	193	194	195	196	197	198	199	200

3. List the numbers that are marked with both an X and a circle.

_____ _____ _____ _____ _____ _____ _____ _____ _____ _____ _____

Math Boxes 1.7

1. Write 5 names in the 25-box.

2. Fill in the missing numbers.

	352		
		373	

3. Write the number that is 100 more.

16 ________

104 ________

950 ________

Write the number that is 100 less.

249 ________

527 ________

4. Count back by 4s.

104, ______, ______, ______,

88, ______, ______, ______,

______, ______, ______, ______,

______, ______, ______, ______

5. Draw hands on the clock to show 6:45.

6. Add.

$2 + 8 =$ ________

$5 + 3 =$ ________

________ $= 6 + 7$

________ $= 7 + 9$

$5 + 8 =$ ________

Date Time

Using a Calculator

Math Message

Use your calculator.

1. Sharon read the first 115 pages of her book last week. She read the rest of the book this week. If she read 86 pages this week, how many pages long is her book?

   Answer: Her book is ________ pages long.

   Number model: ______________________________

2. The paper clip was invented in 1868. The stapler was invented in 1900. How many years after the paper clip was the stapler invented?

   Answer: The stapler was invented ________ years later.

   Number model: ______________________________

3. 28 + 64 + 39 = ________

4. 2,648 − 1,576 = ________

Calculator Practice

Use your calculator.

5. Begin at 25. Count up by 6s. Record your counts below.

   25 ___ ___ ___ ___ ___ ___ ___ ___ ___ ___ ___ ___

6. Begin at 90. Count back by 9s.

   90 ___ ___ ___ ___ ___ ___ ___ ___ ___

Solve the calculator puzzles.

7.

Enter	Change to	How?
42	92	________
61	11	________
136	216	________
78	108	________
108	88	________

8.

Enter	Change to	How?
362	862	________
722	3,722	________
1,604	804	________
9,364	9,964	________

Date Time

Math Boxes 1.8

1. What is today's date?

____

What will be the date in 11 days?

____

What will be the date in 2 weeks?

____

2. 1,942

What value does the 4 have? ________

What value does the 9 have? ________

What value does the 1 have? ________

What value does the 2 have? ________

3. Use Ⓠ, Ⓓ, Ⓝ, and Ⓟ.

Show $1.48 in two ways.

4. Find the difference between

74 and 24 ________

48 and 35 ________

60 and 39 ________

26 and 15 ________

5. Complete the bar graph.

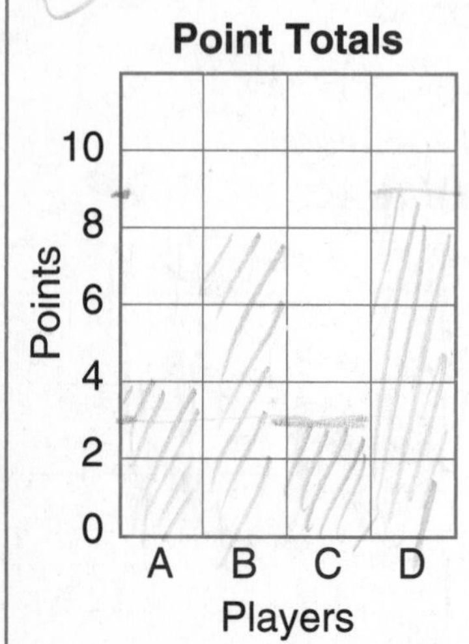

Player A scores 4 points.

Player B scores 8 points.

Player C scores 3 points.

Player D scores 9 points.

6. Add.

$9 + 5 =$ ________

$3 + 7 =$ ________

$5 + 6 =$ ________

________ $= 6 + 8$

________ $= 9 + 3$

Date Time

Using Coins

Math Message

1. You buy a carton of juice for 65 cents. Show two ways to pay for it with exact change. Draw Ⓟs to show pennies, Ⓝs to show nickels, Ⓓs to show dimes, and Ⓠs to show quarters.

   a. ______________________ **b.** ______________________

Write each of the following amounts in dollars-and-cents notation.
The first one is done for you.

Example

three dimes and one nickel $0.35

2. five dimes and seven pennies ________

3. fourteen dimes ________

4. two quarters and four pennies ________

5. three dollars and one nickel and three pennies ________

6. seven dollars and eight dimes ________

Write =, <, or >.

7. three quarters ______ three dimes

8. ten dimes ______ one dollar

9. $0.67 ______ seven dimes

10. $1.18 ______ ⓆⓆⓆⓆⓆ

11. ⒹⒹⓃⓃⓃⓅⓅ ______ ⓆⓃⓅ

12. $2.05 ______ $2.50

Remember

= means *is equal to*

< means *is less than*

> means *is greater than*

Date Time

Using Coins (cont.)

13. Circle the digit that represents dimes.

$17.63

14. Circle the digit that represents cents.

$18.38

15. Circle the digit that represents dimes.

35¢

16. Jean wants to buy a carton of milk for 35¢.
How much change will she get from 2 quarters? ______________

Use Ⓠ, Ⓓ, Ⓝ, and Ⓟ to show her change in two ways.

____

Challenge

Use the Vending Machine Poster on *Student Reference Book,* page 236.

17. Marcy wants to get a strawberry yogurt drink and a chocolate milk from the vending machine. She has only dollar bills.

a. If the Exact Change light is on, can she buy what she wants? __________

b. If the Exact Change is off, how many dollar bills will she put in the machine? __________

How much change will she get? __________

Date Time

Math Boxes 1.9

1. Write 5 names in the 75-box.

75

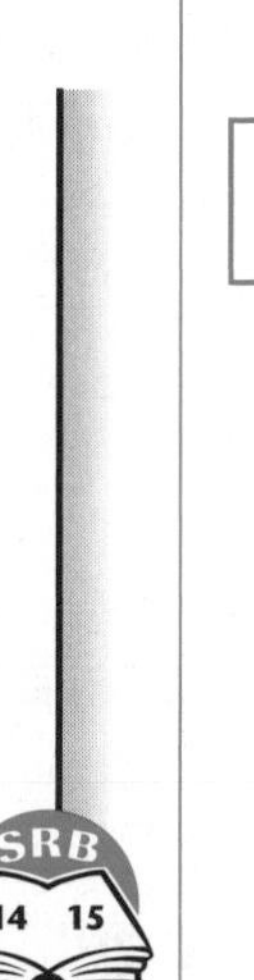

2. Fill in the missing numbers.

	632			
			644	

3. What is 10 more?

614 ________

994 ________

2,462 ________

What is 100 more?

237 ________

3,965 ________

4. Count back.

1,011, 1,010, ________,

________, ________, ________,

________, ________, ________,

________, ________, ________,

________, ________

5. What time does the clock show?

____

What time will it be in 30 minutes? ________

6. Add.

$3 + 6 =$ ________

________ $= 5 + 7$

$8 + 6 =$ ________

$9 + 9 =$ ________

$6 + 4 =$ ________

Date Time

A Shopping Trip

Use the Stationery Store Poster on *Student Reference Book,* page 238.

1. List the items you are buying in the space below. You must buy at least 3 items. You can buy 2 of the same item, but list it twice.

Item	Sale Price
________	________
________	________
________	________

2. Estimate how many dollar bills you will need to give the shopkeeper to pay for your items. ________ dollar bills

3. Give the shopkeeper the dollar bills.

4. The shopkeeper calculates the total cost using a calculator. You owe $________.

5. The shopkeeper calculates the change you should be getting. $________

6. Use Ⓟ, Ⓝ, Ⓓ, Ⓠ, and [$1] to show the change you got from the shopkeeper. ________

Challenge

7. Henry buys one pack of batteries and a box of crayons. How much money does he save buying them on sale instead of paying the regular price?

	Regular Price	Sale Price	Difference
batteries	$____.____	$____.____	Regular total $____.____
crayons	$____.____	$____.____	Sale total $____.____
Total Cost	$____.____	$____.____	**Amount Saved** $____.____

Coin Collections

Get your coin collection or grab a handful of coins from the classroom collection. Complete the problems below.

1. Count each kind of coin. Give a total value for each type of coin.

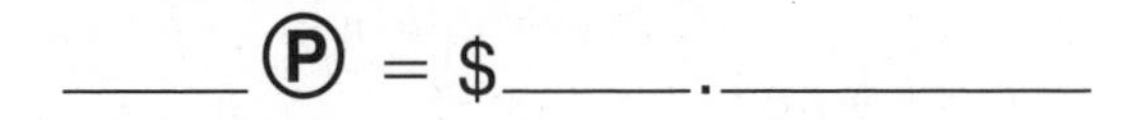

_____ Ⓝ = $_____.__________

_____ Ⓓ = $_____.__________

_____ Ⓠ = $_____.__________

2. What is the total value of all the coins? You may use a calculator.

Total value = $_____.__________

3. In the space below, draw a picture of your total. Use as few [$1], Ⓠ, Ⓓ, Ⓝ, and Ⓟ as possible.

Challenge

4. Explain how you would enter your total amount on the calculator.

____

____

____

5. Explain how you would go up to the next dollar amount without clearing your calculator. (*Hint:* A dollar amount is $1.00, $2.00, $3.00, and so on.)

____

____

____

Date Time

Math Boxes 1.10

1. Use addition or subtraction to complete these problems on your calculator.

Enter	Change to	How?
894	2,894	______
366	66	______
27,581	28,581	______
3,775	3,175	______

SRB 18 19

2. In the number 38,642

the 4 means 40

the 8 means ______

the 6 means ______

the 3 means ______

3. Draw the bills and coins in two ways. $2.43

4. Find the difference between

87 and 37 ______

72 and 55 ______

90 and 49 ______

47 and 26 ______

5. Write <, >, or =.

69 < 96

101 < 110

2Ⓠ = 5Ⓓ

1,000 > 999

6. Complete the bar graph.

Book Club Totals

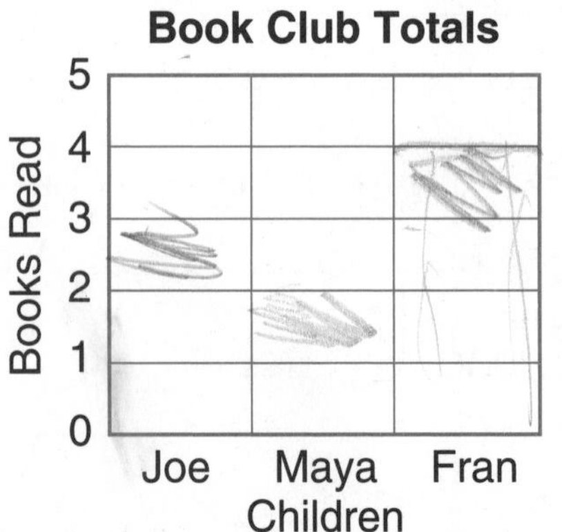

Joe read 3 books.

Maya read 2 books.

Fran read 4 books.

Total books read: ______ (unit)

Date Time

Frames and Arrows

Math Message

Find the pattern. Fill in the missing numbers.

1. 37, 40, 43, _____, _____, _____
2. 27, 25, _____, 21, _____, _____
3. _____, 11, 15, _____, 23, _____
4. _____, _____, 36, 33, _____, 27

Frames and Arrows

5.

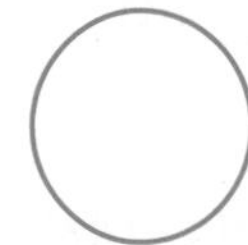

6.

2  8  64

7.

15

8.

30

20

10 5

9. Make up one of your own.

Date Time

Patterns

Complete the number-grid puzzles.

1. 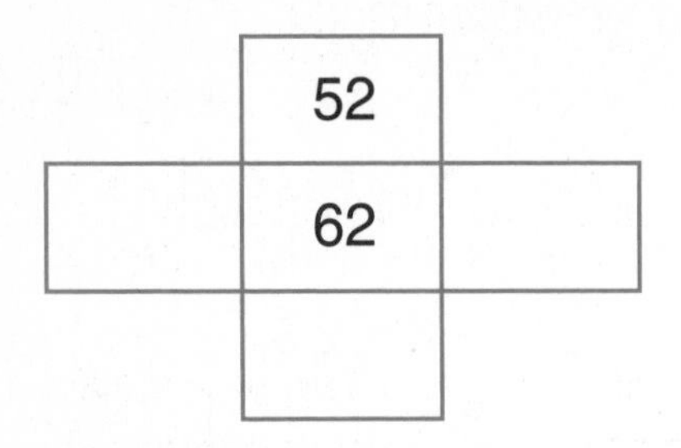

2. 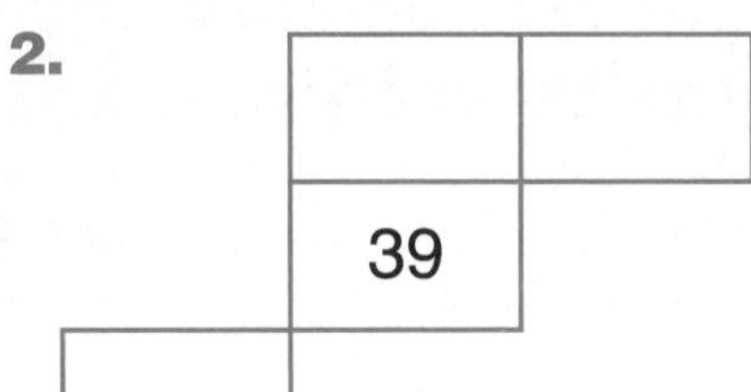

3. 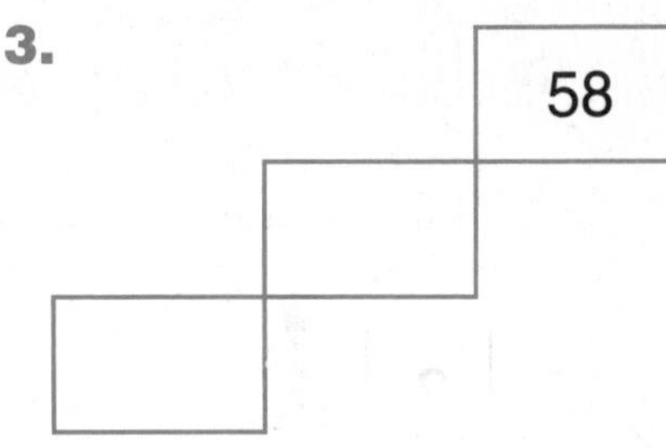

4. Draw dots to show what comes next.

5. Janie owns a magic calculator. When someone enters a number and then presses the (=) key, it changes the number. Here is what happened:
   - Tom entered 15. He pressed (=) and the calculator showed 5.
   - Mary entered 12. She pressed (=) and the calculator showed 2.
   - Regina entered 27. She pressed (=) and the calculator showed 17.

6. What do you think the calculator will show if Janie enters 109 and (=)? ______

7. Explain how you know. ______

Challenge

8. The numbers below have a pattern. Fill in the missing numbers.
   Be careful: The same thing does not always happen each time.

   4, 14, 24, 22, 32, 42, 40, 50, 60, 58, ______, ______, ______

9. Describe the pattern. ______

______

______

Date Time

Tic-Tac-Toe Addition

Draw a line through any three numbers whose sum is the target number in the square. The numbers may be in a row, in a column, or on a diagonal. Draw more than one line for each sum.

8

5	2	1
1	3	7
6	2	0

14

3	4	7
1	8	6
5	1	3

18

6	4	9
5	8	5
7	6	4

20

12	1	9
4	3	6
8	7	5

Think of some other Tic-Tac-Toe puzzles and write them below.

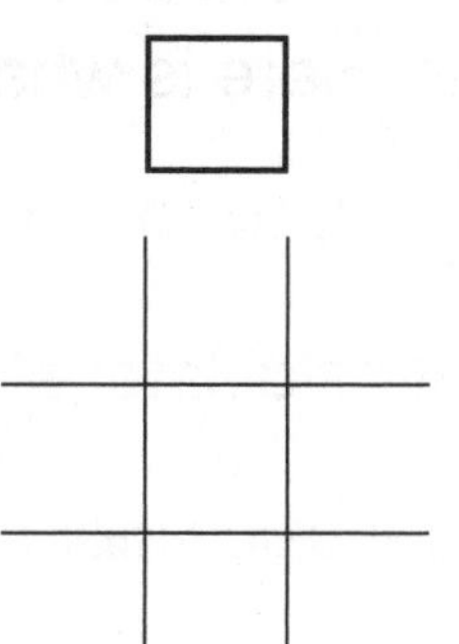

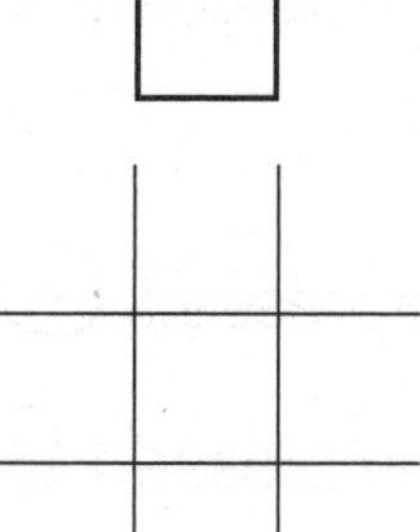

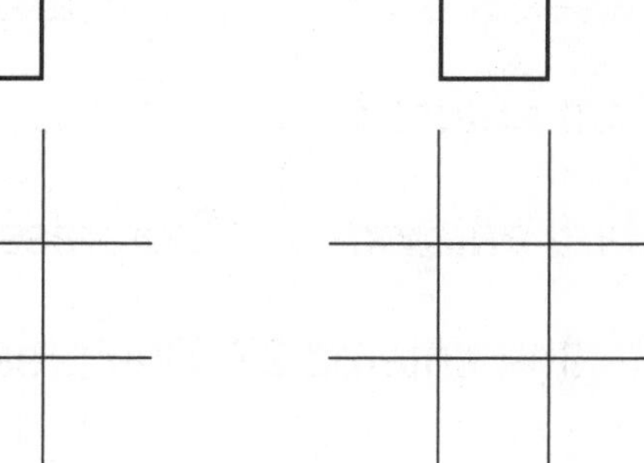

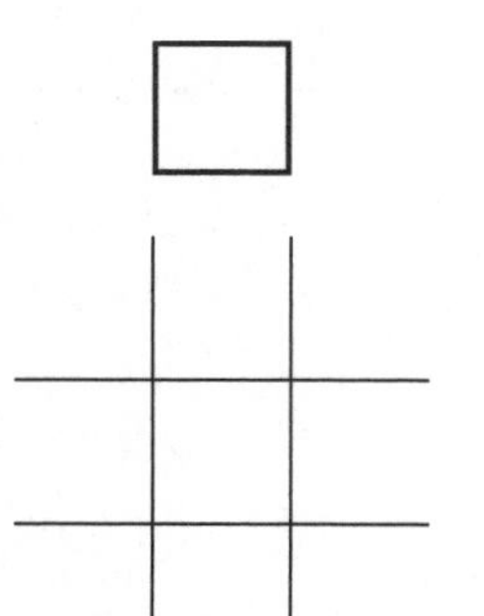

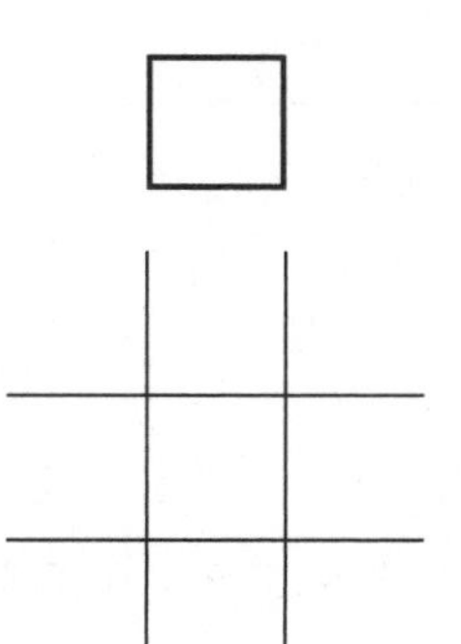

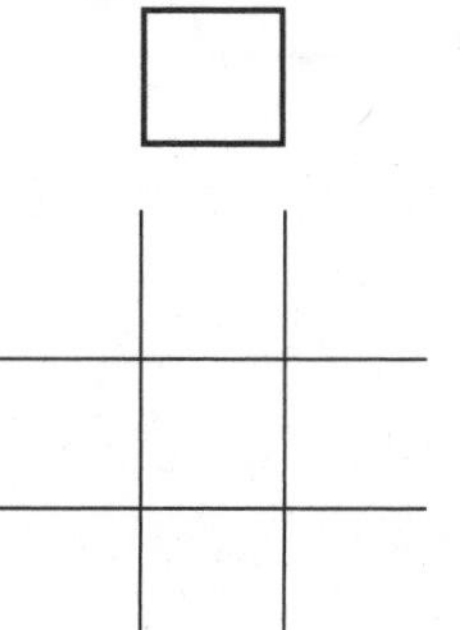

Date Time

Math Boxes 1.11

1. Write 5 names in the 100-box.

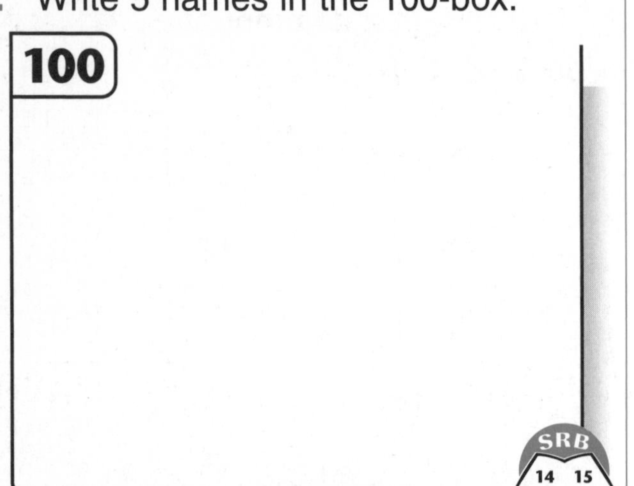

2. Fill in the missing numbers.

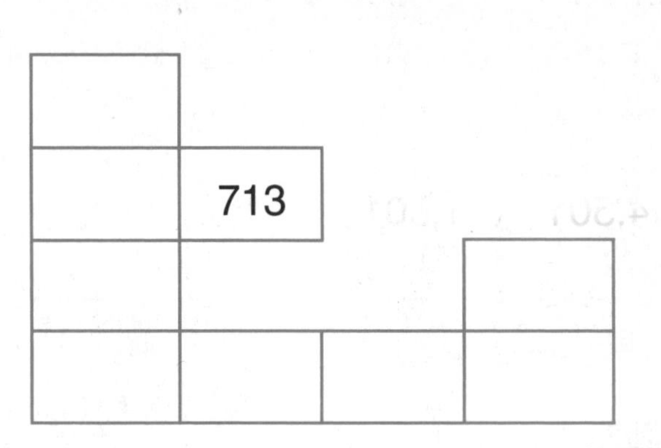

3. Write the number that is

	10 less	100 less	1,000 less
4,321	______	______	______
6,942	______	______	______
7,011	______	______	______
8,002	______	______	______

SRB 18 19

4. Count back.

13, ______, ______, ______,

______, ______, ______, ______,

______, 4, ______, ______,

______, ______, ______, ______,

______

5. About what time is it?

______

How many minutes until 2:00? ______

6. Add.

$4 + 9 =$ ______

$2 + 6 =$ ______

$8 + 7 =$ ______

______ $= 6 + 6$

______ $= 9 + 8$

Date Time

Math Boxes 1.12

1. Use addition or subtraction to complete these problems on your calculator.

Enter	Change to	How?
4,501	1,501	________
173	873	________
15,604	16,604	________
9,646	9,346	________

SRB 18 19

2. Write the number that has

4 hundreds

6 thousands

7 ones

2 tens

____

Read it to a partner.

3. I spend $3.25 at the store. I give the cashier a $5.00 bill.

How much change should I get?

____

4. Find the difference between

91 and 21 ________

53 and 15 ________

70 and 29 ________

83 and 57 ________

5.

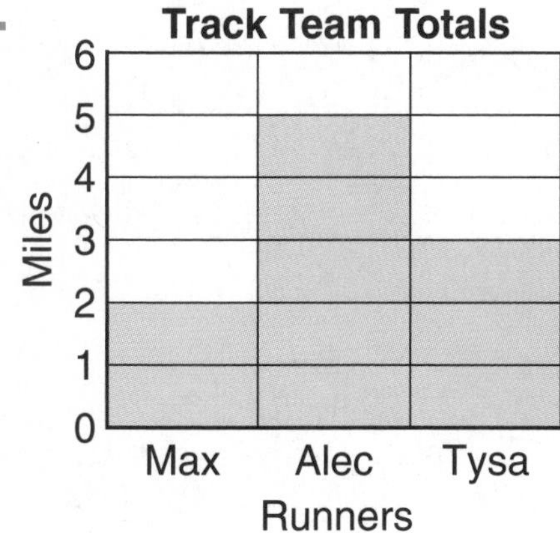

Max ran ______ miles.

Alec ran ______ miles.

Tysa ran ______ miles.

6. Fill in the empty frames.

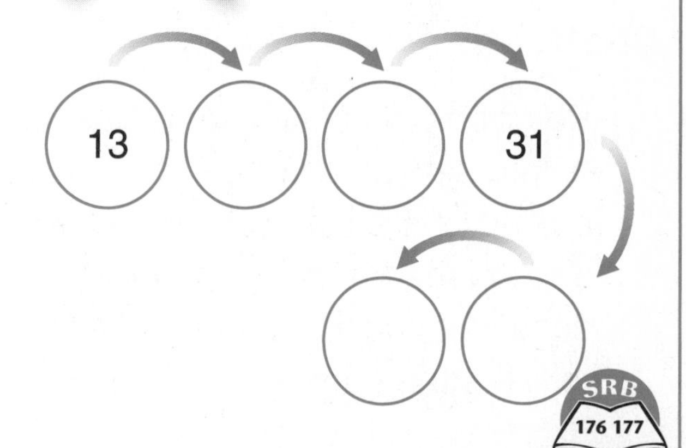

SRB 176 177

Date Time

Finding Elapsed Times

Write the time shown on the first two clocks below. For the third clock, draw the hands to match the time.

1. 

____

2. 

____

3. 

9:15

4. Megan leaves to go swimming at 4:05 and returns at 5:25. How long has she been gone? ______________

5. Robert rides his bike 37 miles. He rides from 10:15 A.M. until 3:50 P.M. How long does it take him to ride 37 miles? ______________

6. Joy leaves for school at the time shown on the first clock. She returns home at the time shown on the second clock. How long is Joy away from home?

____

Challenge

7. Peter baked cookies for a class party. He baked several different kinds. He began baking at the time shown on the first clock and finished at the time shown on the second clock. How long did it take Peter to bake the cookies?

____

Date Time

Math Boxes 1.13

1. Complete the fact family.

   $6 + 7 =$ 13

   $7 +$ 6 $= 13$

   $13 - 6 =$ 7

   13 $- 7 = 6$

2. Lara brought 14 candies to school. She gave away 7 during recess. How many candies does she have now?

   7 candies

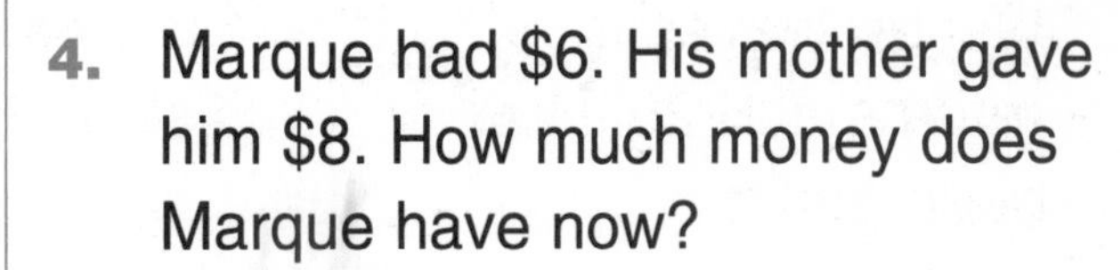

3. Allison swam 16 laps in the pool. Melodia swam 9. How many more laps did Allison swim than Melodia?

   7 laps

4. Marque had \$6. His mother gave him \$8. How much money does Marque have now?

   \$14

5. Andre scored 7 points. Tina scored 5 points. How many points did they score altogether?

   12 points

SRB 188 189

6. Add.

   $0 + 7 =$ 7

   $5 + 1 =$ 6

   $3 + 3 =$ 6

   11 $= 4 + 7$

   15 $= 9 + 6$

Date Time

Math Boxes 2.1

1. Write 5 names in the 120-box.

120

SRB 14 15

2. In the number 76,135

the 1 means 100

the 7 means ______

the 6 means ______

the 3 means ______

3. Show $21.62 in two ways.

4. 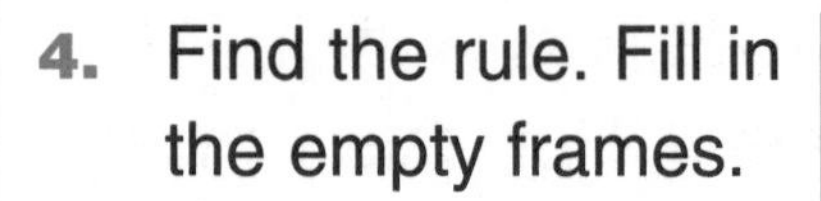
Find the rule. Fill in the empty frames.

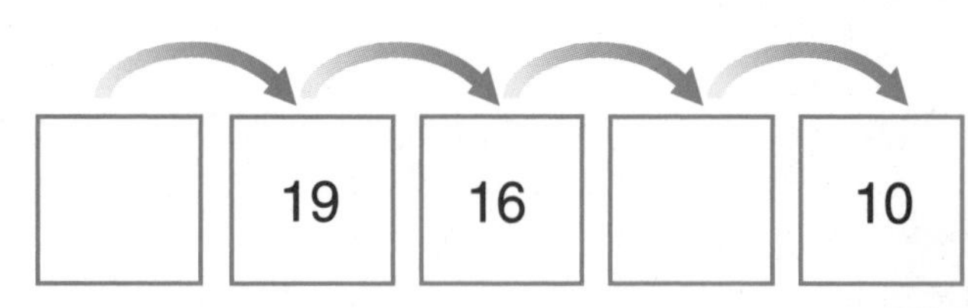

5. Write <, >, or =.

42,617 ☐ 42,429

6,589 ☐ 6,859

1,069 ☐ 10,691

Make up your own.

______ ☐ ______

6. Find the difference between

84 and 14 ______

68 and 25 ______

50 and 16 ______

66 and 42 ______

Date Time

Fact Families and Number Families

Complete the Fact Triangles. Write the fact families.

1.

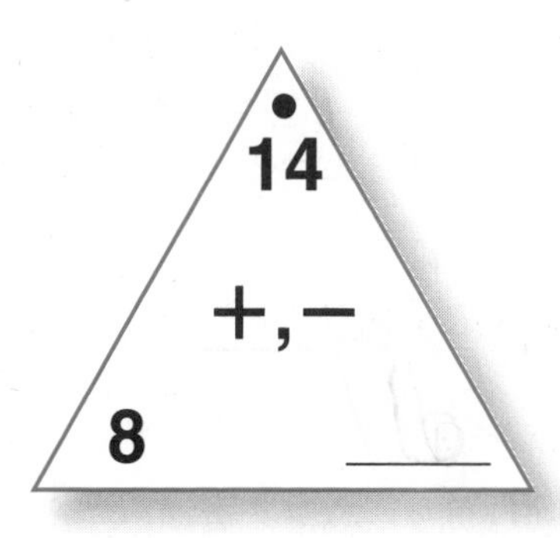

___ + ___ = ___
___ + ___ = ___
___ − ___ = ___
___ − ___ = ___

2.

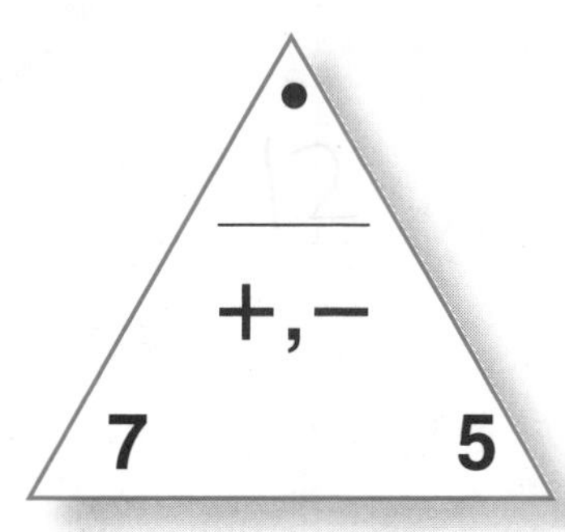

___ = ___ + ___
___ = ___ + ___
___ = ___ − ___
___ = ___ − ___

3.

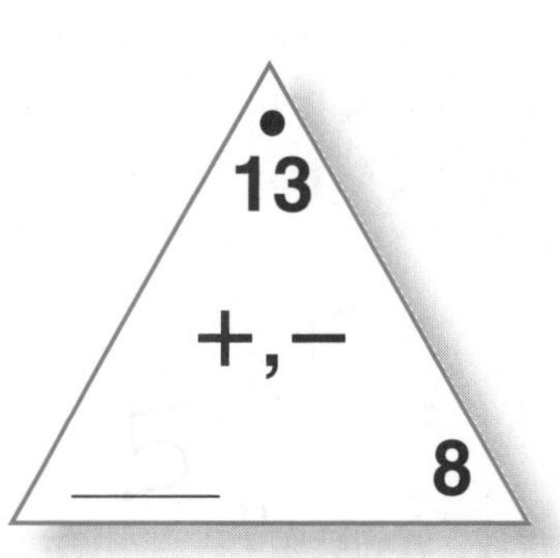

___ + ___ = ___
___ + ___ = ___
___ − ___ = ___
___ − ___ = ___

Complete the number triangles. Write the number families.

4.

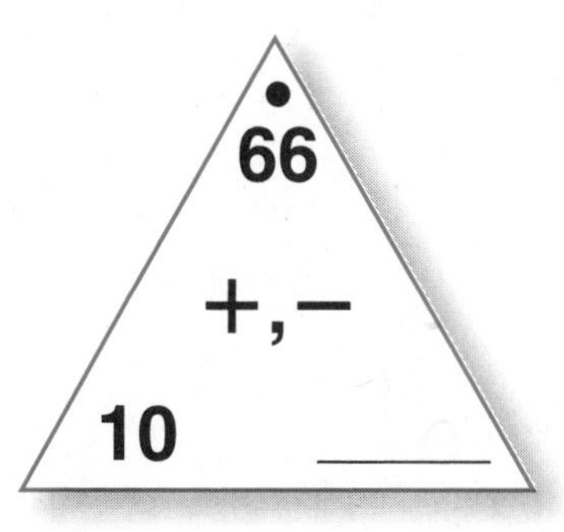

___ = ___ + ___
___ = ___ + ___
___ = ___ − ___
___ = ___ − ___

5.

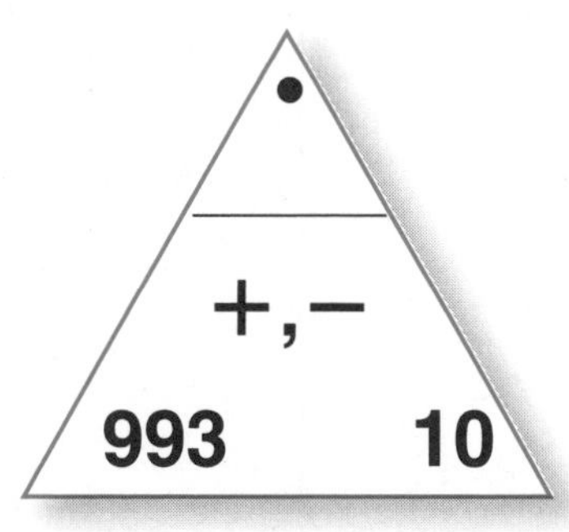

___ + ___ = ___
___ + ___ = ___
___ − ___ = ___
___ − ___ = ___

6.

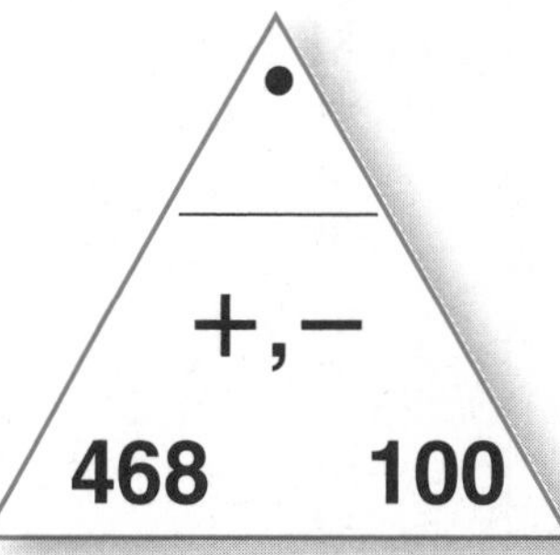

___ = ___ + ___
___ = ___ + ___
___ = ___ − ___
___ = ___ − ___

Date Time

Name-Collection Boxes

1. Three names do not belong. Mark them with a big **X**.

100

1,680 − 1,580

25 +25 + 25

30 + 70

$\begin{array}{r} 80 \\ +\ 30 \\ \hline \end{array}$

$\begin{array}{r} 1,000 \\ -\ 100 \\ \hline \end{array}$

$\begin{array}{r} 63 \\ +\ 37 \\ \hline \end{array}$

$\begin{array}{r} 9,999 \\ -\ 9,899 \\ \hline \end{array}$

2 fifties

48 + 52

2. Write at least 10 names for 40.

40

3. Write at least 10 names for 200.

200

4. Write at least 10 names for 1,000.

1,000

Use with Lesson 2.2.

Date Time

Using Basic Facts to Solve Fact Extensions

Fill in the unit box.

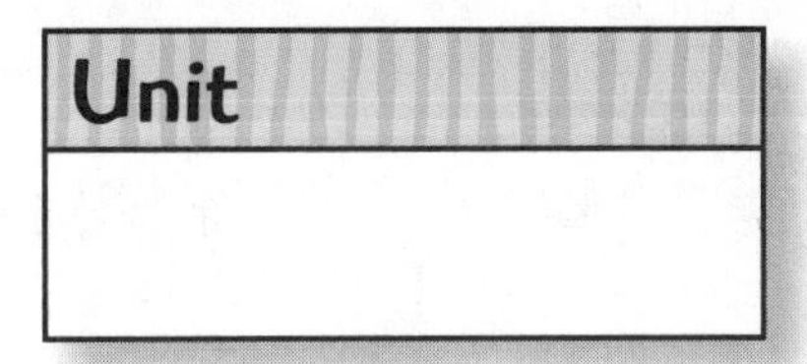

Complete the fact extensions.

1. ______ = 12 − 7
______ = 120 − 70
______ = 1,200 − 700

2. 8 + 3 = ______
80 + 30 = ______
800 + 300 = ______

3. ______ = 7 + 6
______ = 70 + 60
______ = 700 + 600

Complete the fact extensions.

4. ______ = 6 + 8
______ = 16 + 8
______ = 56 + 8

5. 14 − 9 = ______
24 − 9 = ______
54 − 9 = ______

6. ______ = 17 − 11
______ = 27 − 11
______ = 47 − 11

Use addition or subtraction to complete these problems on your calculator.

7.

Enter	Change to	How?
33	40	______
80	73	______
80	23	______

8.

Enter	Change to	How?
430	500	______
700	640	______
1,000	400	______

9. Why is it important to know the basic addition and subtraction facts?

____

____

____

Date Time

Math Boxes 2.2

1. I spent $7.88 at the store. I gave the cashier a $10 bill. How much change should I get back?

$______

2. Write the +, − fact family for 8, 7, and 15.

___ + ___ = ___

___ + ___ = ___

___ − ___ = ___

___ − ___ = ___

3. Use your calculator to find the total.

4 [$1] = $___.___

3 (Q) = $_0_.___

5 (D) = $___.___

7 (N) = $___.___

2 (P) = $___.___

Total $___.___

4. What time is it?

______

What time will it be in 20 minutes?

______

How many minutes until 5:15?

______

5. Put these numbers in order from smallest to largest.

1,060 ______

1,600 ______

1,006 ______

6,001 ______

6. Fill in the missing numbers.

	1,073		
		1,104	

SRB 7 8

Math Boxes 2.3

1. Write the number that is

	10 less	100 more	1,000 more
368	______	______	______
4,789	______	______	______
40,870	______	______	______
1,999	______	______	______

2. Complete the fact extensions.

Unit

$13 = 6 + 7$

______ $= 16 + 7$

______ $= 26 + 7$

______ $= 106 + 7$

______ $= 136 + 7$

3. Show \$6.62 in two other ways.

\$5

Ⓠ Ⓠ

Ⓠ Ⓠ

Ⓠ Ⓠ

Ⓝ Ⓝ

Ⓟ Ⓟ

4. Fill in the empty frames.

Rule
+100

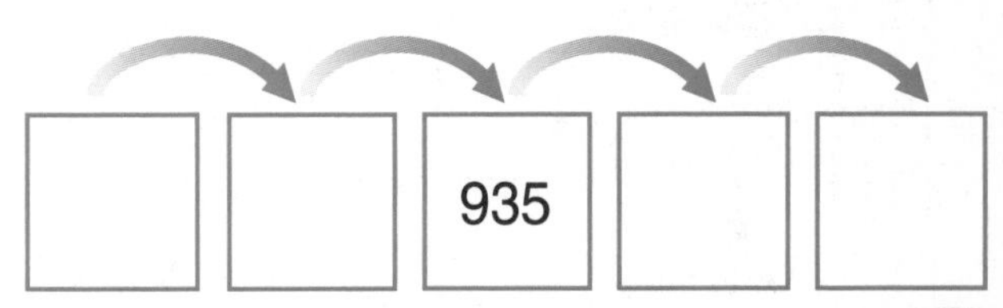

		935		

5. Fill in <, >, or =.

49 ☐ 495

3,000 ☐ 300

69 hundreds 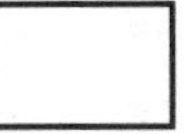 69 thousands

6. Fill in the missing numbers.

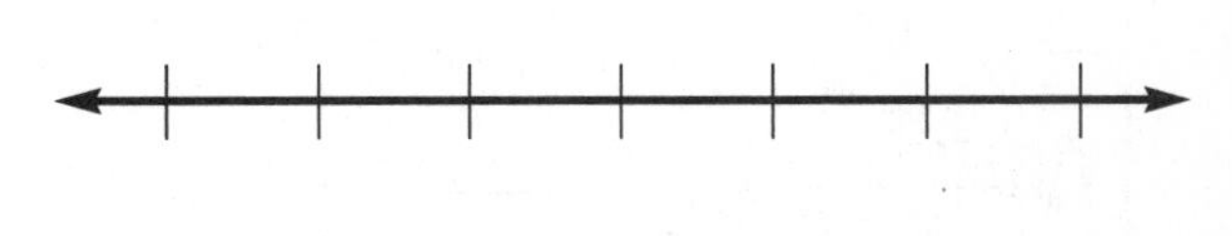

___ ___ 8 ___ ___ 20 ___

SRB 10

"What's My Rule?"

Fill in the blanks.

Unit
stickers

1.

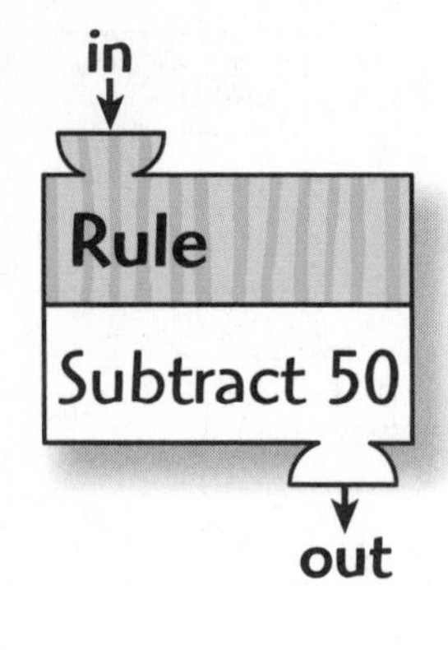

in	out
100	
120	
70	
150	
200	

2.

in
Rule
Subtract 30
out

in	out
	30
	50
	100
	200
	0

3.

in
Rule
Add 8
out

in	out
	13
	23
	43
	73
	93

4.

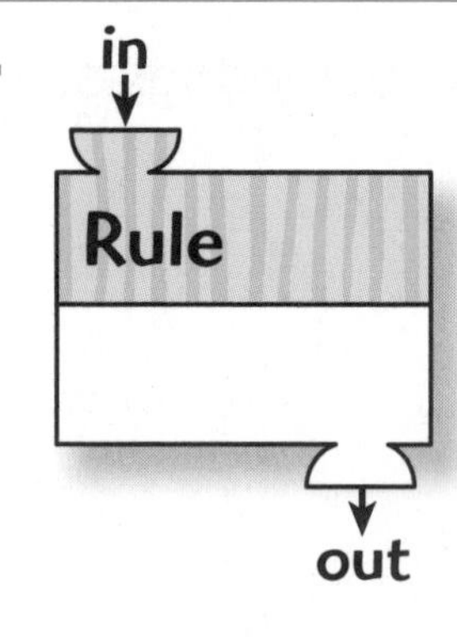

in	out
14	23
34	43
44	53
64	73
94	103

5.

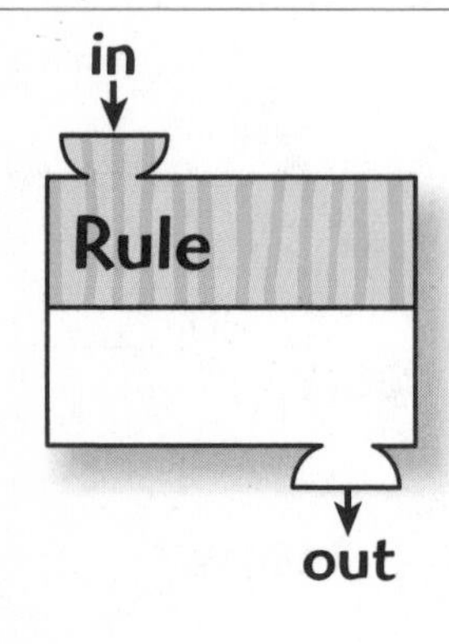

in	out
35	20
	45
20	
50	35
46	

6.

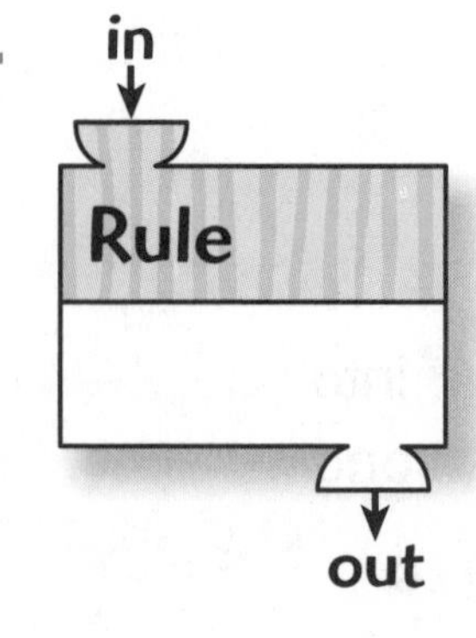

in	out
6	13
9	
5	
4	11
	18

Date Time

Fact Families and Number Families

1. Complete the Fact Triangles. Write the fact families.

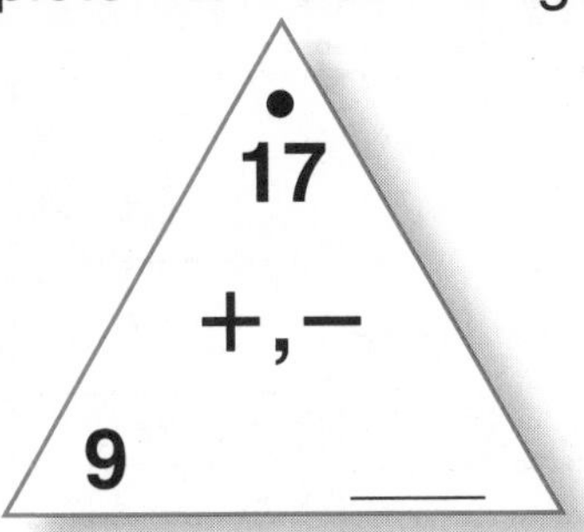

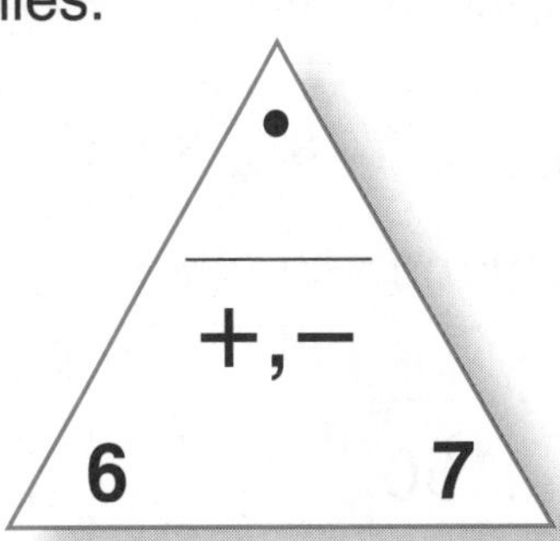

______ + ______ = ______

______ + ______ = ______

______ − ______ = ______

______ − ______ = ______

______ + ______ = ______

______ + ______ = ______

______ − ______ = ______

______ − ______ = ______

2. Complete the number triangles. Write the number families.

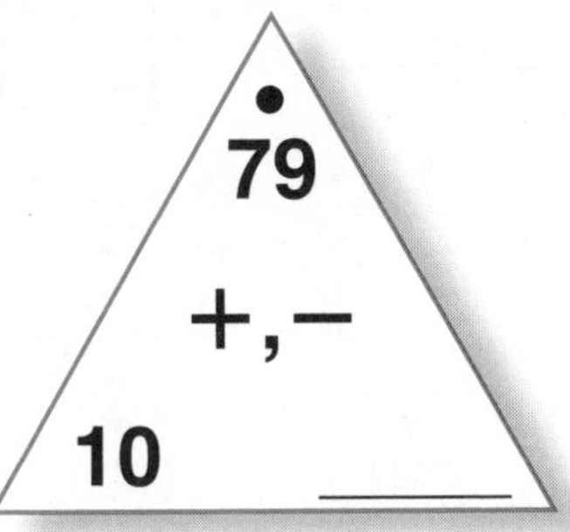

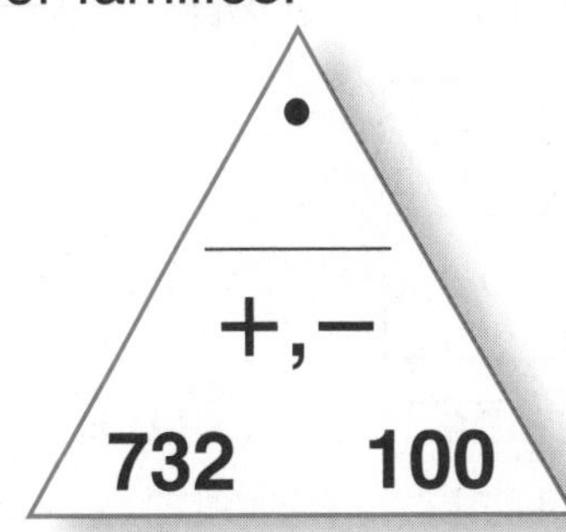

______ + ______ = ______

______ + ______ = ______

______ − ______ = ______

______ − ______ = ______

______ + ______ = ______

______ + ______ = ______

______ − ______ = ______

______ − ______ = ______

Enter the first number into your calculator. Use addition or subtraction to change it to the second number. Then tell what you did.

	Enter	Change to	How?
3.	54	60	______
5.	90	81	______

	Enter	Change to	How?
4.	230	300	______
6.	800	720	______

Date Time

Number Stories: Animal Clutches

For each number story, write the numbers you know in the parts-and-total diagram. Write ? for the number you want to find. Solve the problem and write a number model.

1. Two pythons laid clutches of eggs. One clutch had 36 eggs. The other had 23 eggs. That was how many eggs in all?

   Answer the question: ______________ (unit)

   Number model: ______________

   Check: Does my answer make sense?

Total	
Part	Part

---

2. A queen termite laid about 6,000 eggs on Monday and about 7,000 eggs on Tuesday. About how many eggs did she lay in all?

   Answer the question: ______________ (unit)

   Number model: ______________

   Check: Does my answer make sense?

Total	
Part	Part

---

3. Two agama lizards laid clutches of eggs. One clutch had 19 eggs. The other had 22 eggs. In all, how many eggs were laid?

   Answer the question: ______________ (unit)

   Number model: ______________

   Check: Does my answer make sense?

Total	
Part	Part

---

4. Two clutches of Mississippi alligator eggs were found. Each clutch had 47 eggs. What was the total number of eggs found?

   Answer the question: ______________ (unit)

   Number model: ______________

   Check: Does my answer make sense?

Total	
Part	Part

Date Time

Number Stories: Animal Clutches (cont.)

5. Three ostriches laid clutches of eggs. The first clutch had 15 eggs, the second had 9 eggs, and the third had 10 eggs. That was how many eggs in all?

Total		
Part	Part	Part

Answer the question: ______________ (unit)

Number model: ______________

Check: Does my answer make sense?

Challenge

6. An alligator clutch had 60 eggs. Only 12 eggs hatched. How many eggs did not hatch?

Total	
Part	Part

Answer the question: ______________ (unit)

Number model: ______________

Check: Does my answer make sense?

7. Scientists say a green turtle can lay about 1,800 eggs in a lifetime. But only about 400 eggs hatch. About how many eggs do not hatch?

Total	
Part	Part

Answer the question: ______________ (unit)

Number model: ______________

Check: Does my answer make sense?

8. On a separate sheet of paper, make up and solve a story using the Animal Clutches poster on pages 242 and 243 in your *Student Reference Book*.

Total	
Part	Part

Answer the question: ______________ (unit)

Number model: ______________

Check: Does my answer make sense?

"What's My Rule?"

Fill in the blanks.

1. in → Rule: Add 20 minutes → out

in	out
1:00	
2:05	
4:15	
7:45	
8:51	

2. in → Rule: Subtract 10 minutes → out

in	out
	2:00
	3:15
	6:35
	7:42
	9:55

3. in → Rule → out

in	out
2:00	2:50
3:15	4:05
5:30	6:20
	7:55
8:45	

4. in → Rule: Add 25¢ → out

in	out
10¢	
20¢	
25¢	
83¢	
$1.00	

5. in → Rule: Subtract 10¢ → out

in	out
	20¢
	45¢
	50¢
	63¢
	$1.00

6.

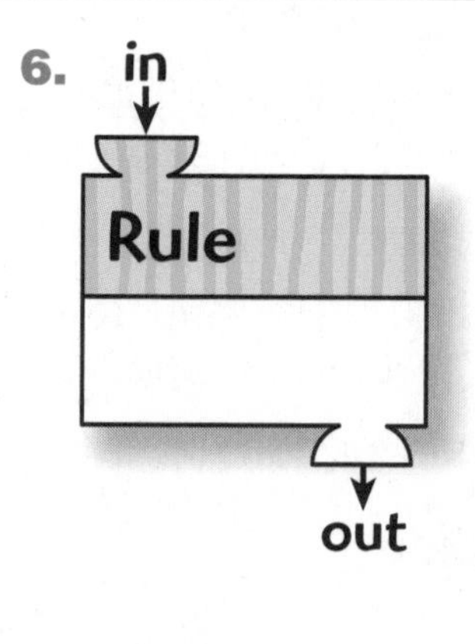

in	out
10¢	26¢
25¢	41¢
$1.20	$1.36
80¢	
	99¢

Date Time

Math Boxes 2.4

1. I had a $10 bill. I bought $3.92 worth of candy. How much change should I get?

____

2. Complete the Fact Triangle. Write the fact family.

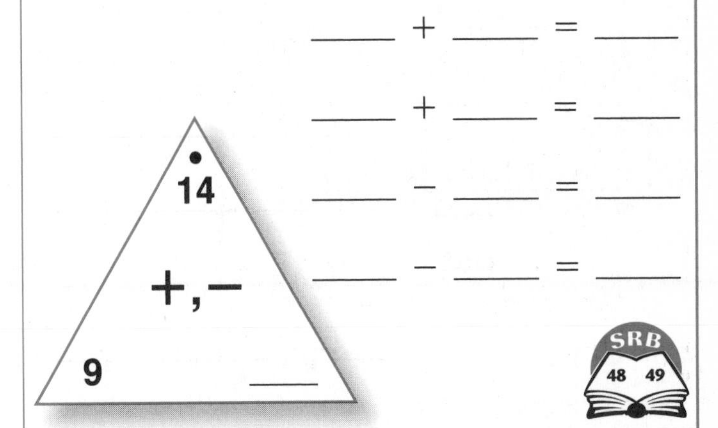

3. Use a calculator to find the total.

2 [$1] = $_______

1 (Q) = $_______

3 (D) = $_______

8 (N) = $_______

6 (P) = $_______

Total $_______

4. "What's My Rule?"

in	out
14	
24	
39	
	42
	65

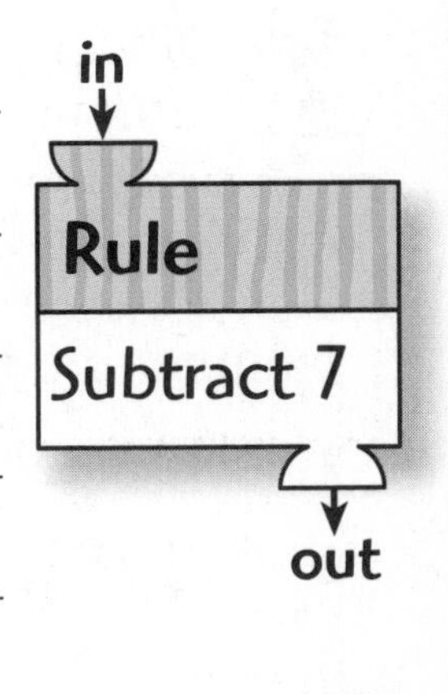

5. Use addition or subtraction to complete these problems on your calculator.

Enter	Change to	How?
4,501	1,501	_______
173	873	_______
15,604	16,604	_______
9,646	9,346	_______

SRB 18 19

6. Find the difference between

71 and 41 _____

93 and 45 _____

60 and 22 _____

87 and 54 _____

Date Time

Number Stories: Change-to-More and Change-to-Less

For each number story, write the numbers you know in the change diagram. Write ? for the number you want to find. Then solve the problem. Write the answer and a number model.

Unit
dollars

1. David had $22 in his bank account. For his birthday, his grandmother deposited $25 for him. How much money is in his bank account now?

Start	Change	End

   Answer the question: ____________________

   Number model: ____________________

   Check: Does my answer make sense?

2. Jennifer had $19 in her bank account. After babysitting, she is able to deposit $38. How much money is in her bank account now?

Start	Change	End

   Answer the question: ____________________

   Number model: ____________________

   Check: Does my answer make sense?

3. Omar had $53 in his piggy bank. He used $16 to take his sister to the movies and buy treats. How much money is left in his piggy bank?

Start	Change	End

   Answer the question: ____________________

   Number model: ____________________

   Check: Does my answer make sense?

4. Cleo had $37 in her purse. Then Jillian returned $9 that she had borrowed. How much money does Cleo have now?

Start	Change	End

   Answer the question: ____________________

   Number model: ____________________

   Check: Does my answer make sense?

Number Stories (cont.)

5. Tyler had $30 in his wallet. At lunch he spent $17. How much money does Tyler have now?

Answer the question: ______________________

Number model: ______________________

Check: Does my answer make sense?

---

6. Andre had $61 in his bank account. He withdrew $48 to take on vacation. How much is left in his account?

Start | Change | End

Answer the question: ______________________

Number model: ______________________

Check: Does my answer make sense?

Challenge

7. Trung had $15 in his piggy bank. After his birthday, he has $60 in his bank. How much money did Trung get as birthday presents?

Start | Change | End

Answer the question: ______________________

Number model: ______________________

Check: Does my answer make sense?

---

8. Nikhil had $40 in his wallet when he went to the carnival. When he got home, he had $18. How much did he spend at the carnival?

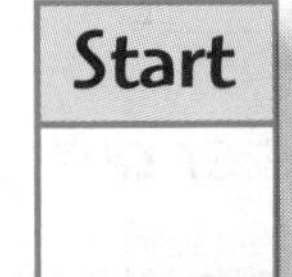

End

Answer the question: ______________________

Number model: ______________________

Check: Does my answer make sense?

Date Time

Parts-and-Total Number Stories

For each number story, write the numbers you know in the parts-and-total diagram. Write ? for the number you want to find. Then solve the problem. Write the answer and a number model.

1. There were 80 people at the concert on Saturday night and 50 people at the concert on Sunday night. Altogether, how many people went to the concert?

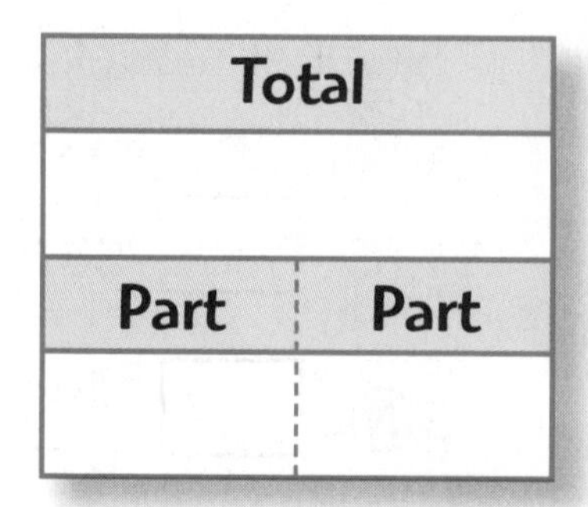

Answer the question: ______________________ (unit)

Number model: ______________________

Check: Does my answer make sense?

---

2. About 800 pieces of mail are lost in the United States every day. About how many pieces of mail are lost in 2 days?

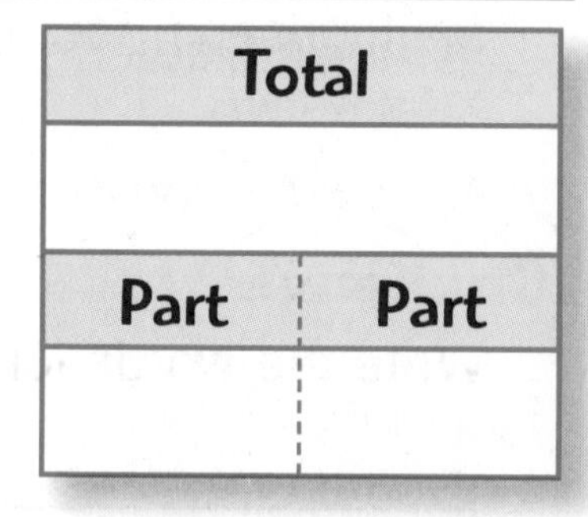

Answer the question: ______________________ (unit)

Number model: ______________________

Check: Does my answer make sense?

---

3. The Ramirez family drove 600 miles during the first week of their vacation and 900 miles during the second week. How many miles did they drive in all?

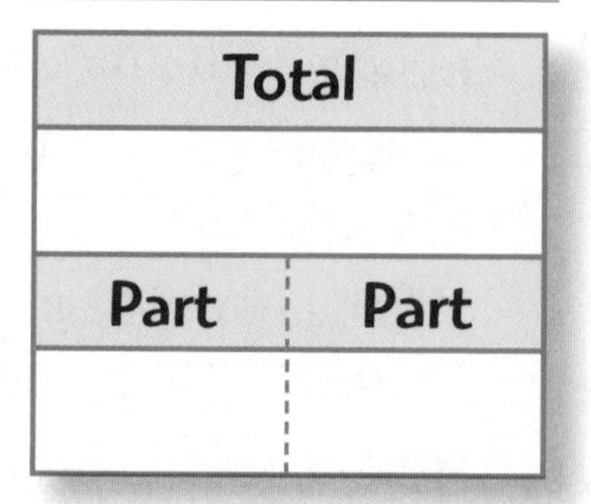

Answer the question: ______________________ (unit)

Number model: ______________________

Check: Does my answer make sense?

Date Time

Math Boxes 2.5

1. Write <, >, or =.

45¢ ☐ $0.45

4Ⓓ ☐ 3Ⓠ

$1.85 ☐ $3.00

5Ⓝ ☐ 2Ⓓ, 1Ⓝ

2. Find the missing sums.

Unit

4 + 5 = ______

______ = 14 + 5

24 + 5 = ______

5 + 44 = ______

3. Write this number:

six thousand, four hundred twenty-two

______

Write the words for 5,931.

______

______

4. The school chorus has 28 second graders and 34 third graders. How many children are in the chorus?

______ children

Total	
Part	Part

5. How many children like grapes?

______

How many children like oranges?

______

Fruit Choice	Number of Children
apples	////
grapes	~~////~~ /
oranges	///
pears	~~////~~ ~~////~~

6. Fill in the empty frames. Use two rules.

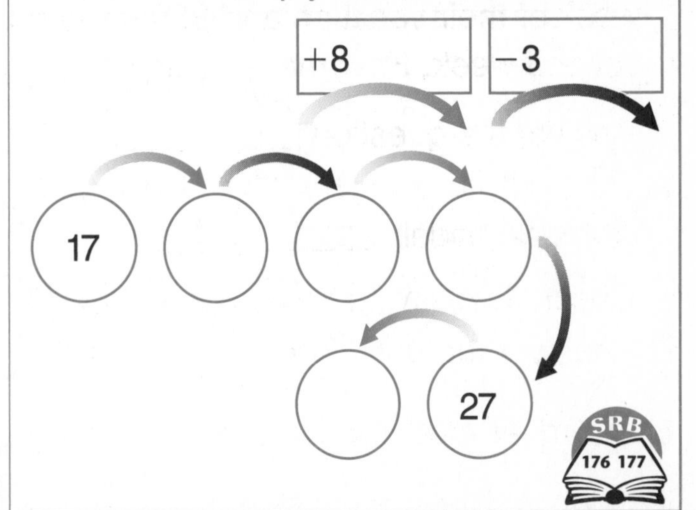

Date Time

Temperature Differences

Use the map on page 244 in the *Student Reference Book* to answer Problems 1–4. Write the numbers you know in the comparison diagram. Write ? for the number you want to find. Then solve the problem. Write the answer and a number model.

1. What is the difference between the normal high and low temperatures for San Francisco?

   Answer the question: ________ °F

   Number model: ____________________

   Check: Does my answer make sense?

Quantity	
Quantity	**Difference**

2. What is the difference between the normal high and low temperatures for Minneapolis?

   Answer the question: ________ °F

   Number model: ____________________

   Check: Does my answer make sense?

Quantity	
Quantity	**Difference**

3. Which city has the *largest* difference between the normal high and low temperatures?

   ______________ What is the difference? ________ °F

4. Which city has the *smallest* difference between the normal high and low temperatures?

   ______________ What is the difference? ________ °F

5. The normal January low in Chicago is 25°F less than the normal spring low of 38°F. What is the normal January low in Chicago?

   Answer the question: ________ °F

   Number model: ____________________

   Check: Does my answer make sense?

Quantity	
Quantity	**Difference**

Date Time

Math Boxes 2.6

1. Write at least 5 names for 1,000.

1,000

SRB 14 15

2. Use 15, 12, and 27. Write the number family.

______

______

______

______

3. 14 dimes = $_____._____

14 nickels = $_____._____

14 quarters = $_____._____

3 quarters and 6 dimes

= $_____._____

4. "What's My Rule?"

in	out
4	
	12
0	
	21

in

Rule

Add 9

out

SRB 179 180

5. Complete the grid.

		9,975	

SRB 7 8

6. Jonah had $52. He bought a CD for $14. How much money does he have now?

______

SRB 186 187

Date Time

The Partial-Sums Addition Method

Make a ballpark estimate first. Write a number model to show your estimate. Next, solve using the partial-sums method and show your work. Then compare your answers with a partner's. If you disagree, use a calculator. If you did a problem incorrectly, work it again.

Unit
miles

Example

	100s	10s	1s
	3	2	9
+	4	1	8
	7	0	0
		3	0
+		1	7
	7	4	7

Ballpark estimate:

300 + 400 = 700

1.

$$\begin{array}{r} 43 \\ +\ 26 \\ \hline \end{array}$$

Ballpark estimate:

______

2.

$$\begin{array}{r} 90 \\ +\ 37 \\ \hline \end{array}$$

Ballpark estimate:

______

3.

$$\begin{array}{r} 172 \\ +\ 109 \\ \hline \end{array}$$

Ballpark estimate:

______

4.

$$\begin{array}{r} 87 \\ +\ 113 \\ \hline \end{array}$$

Ballpark estimate:

______

5.

$$\begin{array}{r} 376 \\ +\ 401 \\ \hline \end{array}$$

Ballpark estimate:

______

Date Time

The Partial-Sums Addition Method (cont.)

6.

$$\begin{array}{r} 751 \\ +\ 757 \\ \hline \end{array}$$

Ballpark estimate:

7.

$$\begin{array}{r} 743 \\ +\ 504 \\ \hline \end{array}$$

Ballpark estimate:

8.

$$\begin{array}{r} 257 \\ +\ 245 \\ \hline \end{array}$$

Ballpark estimate:

9.

$$\begin{array}{r} 298 \\ +\ 419 \\ \hline \end{array}$$

Ballpark estimate:

10.

$$\begin{array}{r} 487 \\ +\ 313 \\ \hline \end{array}$$

Ballpark estimate:

11.

$$\begin{array}{r} 1{,}438 \\ +\ 694 \\ \hline \end{array}$$

Ballpark estimate:

Date Time

Change-to-More and Change-to-Less Number Stories

Write the numbers you know in the change diagram. Write ? for the number you want to find. Then solve the problem. Write the answer and a number model.

1. Nikki had a collection of 35 beanbag animals. She gave 17 of the animals to her sister. How many does she have now?

Start	Change	End

Answer the question: ______________ (unit)

Number model: ______________

Check: Does my answer make sense?

2. Lewis delivered newspapers to 27 houses. Fourteen more houses were added to his route. How many houses does he deliver to now?

Start	Change	End

Answer the question: ______________ (unit)

Number model: ______________

Check: Does my answer make sense?

3. At 5:00 P.M. there were 100 people waiting for the fireworks. By 8:00 P.M. 300 more people had arrived. How many people were waiting then?

Start	Change	End

Answer the question: ______________ (unit)

Number model: ______________

Check: Does my answer make sense?

4. Make up your own change number story.

____

____

Answer the question: ______________ (unit)

Start	Change	End

Number model: ______________

Check: Does my answer make sense?

Math Boxes 2.7

1.

	10 more	100 more	1,000 more
65	______	______	______
410	______	______	______
602	______	______	______
1,543	______	______	______
7,095	______	______	______

2. Fill in the blanks.

34 + ______ = 60

______ = 19 + 21

100 = 50 + ______

70 = ______ − 20

3. I spent $4.13 at the store. I gave the cashier $5.00. How much change should I receive?

______

Draw the fewest number of coins possible to show the change I received.

4. Lily had 33 rings in one box and 29 in another. How many did she have in all? ______ rings

Total	
Part	**Part**

5. Fill in the empty frames. Use two rules.

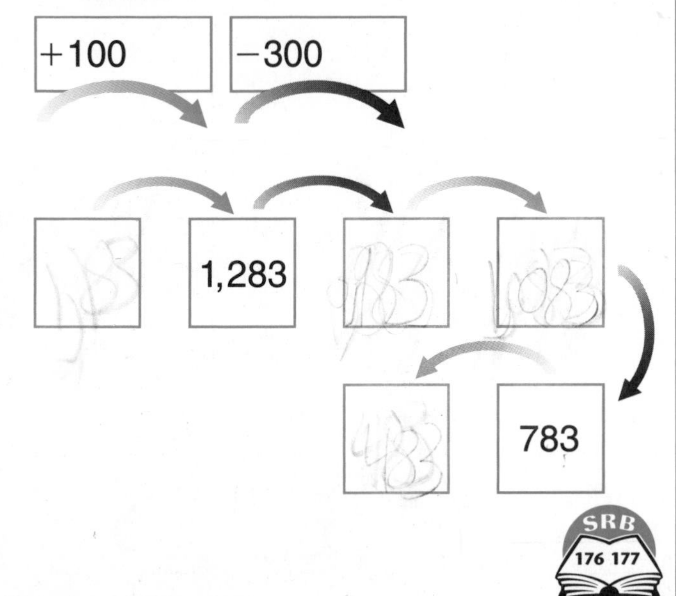

6. Austin read his book for 45 minutes on Monday and for 25 minutes on Tuesday. How many more minutes did he read on Monday?

______ minutes

Quantity	
Quantity	**Difference**

SRB 190

Date Time

The Trade-First Subtraction Method

Solve using the trade-first subtraction method. Show your work. Use a ballpark estimate to check whether your answer makes sense. Write a number model for your estimate. Then compare your answers with a partner's. Use a calculator if you disagree. If you did a problem incorrectly, work it again.

Unit
miles

Example

100s	10s	1s
1	14	
~~2~~	~~4~~	7
−1	8	6
	6	1

Ballpark estimate:

250 − 200 = 50

1.

$$\begin{array}{r} 91 \\ -\ 46 \\ \hline \end{array}$$

Ballpark estimate:

____

2.

$$\begin{array}{r} 63 \\ -\ 38 \\ \hline \end{array}$$

Ballpark estimate:

____

3.

$$\begin{array}{r} 129 \\ -\ 112 \\ \hline \end{array}$$

Ballpark estimate:

____

4.

$$\begin{array}{r} 208 \\ -\ 106 \\ \hline \end{array}$$

Ballpark estimate:

____

5.

$$\begin{array}{r} 213 \\ -\ 206 \\ \hline \end{array}$$

Ballpark estimate:

____

Date Time

The Trade-First Subtraction Method (cont.)

6. 245 − 207 Ballpark estimate: ______	**7.** 283 − 256 Ballpark estimate: ______	**8.** 853 − 606 Ballpark estimate: ______
9. 826 − 172 Ballpark estimate: ______	**10.** 752 − 387 Ballpark estimate: ______	**11.** 640 − 479 Ballpark estimate: ______

Date Time

Addition Strategies

Use any method you like to solve each addition problem. Show your work. Use a ballpark estimate to check whether your answer makes sense. Write a number model for your estimate.

Example	**1.**	**2.**
100s 10s 1s 2 3 8 + 4 4 1 600 70 + 9 679 Ballpark estimate: 240 + 440 = 680	439 + 356 Ballpark estimate: __________	318 + 226 Ballpark estimate: __________
3.	**4.**	**5.**
487 + 258 Ballpark estimate: __________	353 + 187 Ballpark estimate: __________	754 + 668 Ballpark estimate: __________

Date Time

Math Boxes 2.8

1. Put these numbers in order from smallest to largest.

32,764 __________

8,596 __________

32,199 __________

85,096 __________

2. Use 87, 5, and 92. Write 2 addition and 2 subtraction number models.

____

____

____

____

3. Add. Show your work.

$$\begin{array}{r} 27 \\ +\ 48 \\ \hline \end{array} \qquad \begin{array}{r} 152 \\ +\ 394 \\ \hline \end{array}$$

4. "What's My Rule?"

in	out
10	
21	
32	
	60

in

Rule

Add 4

out

SRB 179 180

5. Use your calculator. Write the answers in dollars and cents.

64¢ + \$1.73 = \$_____._____

\$0.85 + 53¢ = \$_____._____

\$2.08 + \$5.01 = \$_____._____

37¢ + 26¢ = \$_____._____

6. Theo had 17 shells in his collection. He found 9 more at the beach. How many shells are in his collection now?

__________ shells

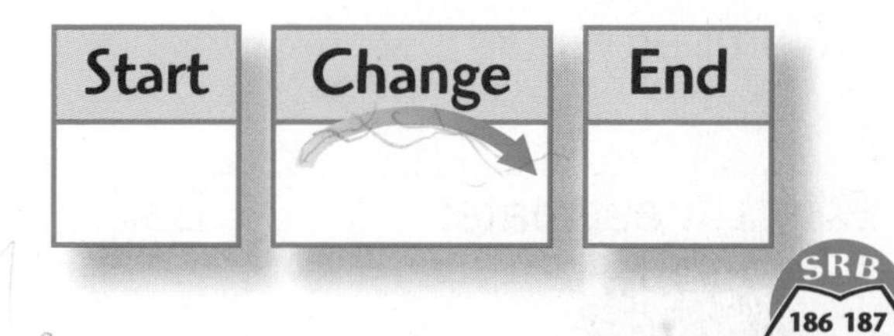

SRB 186 187

Date Time

Number Stories with Three or More Addends

1. José bought milk at 35 cents, apple juice at 55 cents, grape juice at 45 cents, and orange juice at 65 cents. How much money did he spend?

Total			
Part	Part	Part	Part

Answer the question: ______ (unit)

Number model:

____

Check: Does my answer make sense?

2. Michelle drove from Houston, Texas, to Wichita, Kansas. On the first day she drove 245 miles. On the second day she drove 207 miles. On the third day she drove 158 miles and arrived in Wichita. How many miles did she travel in all?

Total		
Part	Part	Part

Answer the question: ______ (unit)

Number model:

____

Check: Does my answer make sense?

3. Zookeepers watched a clutch of 54 python eggs. On the first day, 18 eggs hatched. On the next day, 11 more hatched. How many eggs still had not hatched?

Total		
Part	Part	Part

Answer the question: ______ (unit)

Number model:

____

Check: Does my answer make sense?

Date Time

Number Stories with Three or More Addends (cont.)

4. Carl has $2.50 for juice or milk at lunch. On each of 2 days, he buys grape juice for 45 cents. On the third day, he buys milk for 40 cents. How much money does he have left?

Total			
Part	Part	Part	Part

Answer the question: ______________ (unit)

Number model:

____

Check: Does my answer make sense?

---

5. Janna started to read a 128-page book. She read 13 pages before dinner and 39 pages after dinner. How many pages does she have left?

Total		
Part	Part	Part

Answer the question: ______________ (unit)

Number model:

____

Check: Does my answer make sense?

---

6. The Flores family is driving from Minneapolis, Minnesota, to Bismarck, North Dakota. The distance is 501 miles. They drove 235 miles before lunch. After lunch they drove 150 miles and stopped for a rest. How many more miles will they drive?

Total		
Part	Part	Part

Answer the question: ______________ (unit)

Number model:

____

Check: Does my answer make sense?

Date Time

Subtraction Strategies

Solve each subtraction problem using your own method. Show your work. Use a ballpark estimate to check whether your answer makes sense. Write a number model for your estimate.

Example

100s	10s	1s
1	12	
~~2~~	~~2~~	6
−1	3	4
	9	2

Ballpark estimate:

230 − 130 = 100

1.

$$\begin{array}{r} 93 \\ -\ 47 \\ \hline \end{array}$$

Ballpark estimate:

____

2.

$$\begin{array}{r} 487 \\ -\ 129 \\ \hline \end{array}$$

Ballpark estimate:

____

3.

$$\begin{array}{r} 361 \\ -\ 248 \\ \hline \end{array}$$

Ballpark estimate:

____

4.

$$\begin{array}{r} 724 \\ -\ 396 \\ \hline \end{array}$$

Ballpark estimate:

____

5.

$$\begin{array}{r} 515 \\ -\ 367 \\ \hline \end{array}$$

Ballpark estimate:

____

Date Time

Math Boxes 2.9

1. Fill in the tag. Write at least 5 names for that number.

2. Complete the problems.

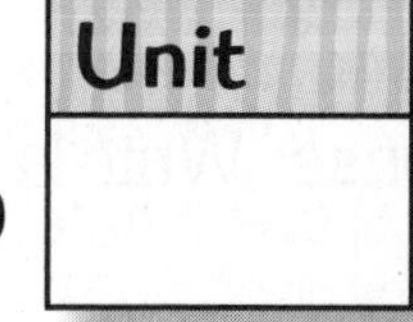

$$\begin{array}{r} 430 \\ + \underline{\quad\quad} \\ 600 \end{array} \qquad \begin{array}{r} 950 \\ + \underline{\quad\quad} \\ 1{,}000 \end{array}$$

$$\begin{array}{r} 1{,}000 \\ - \underline{\ 300} \\ \end{array} \qquad \begin{array}{r} 560 \\ - \underline{\quad\quad} \\ 400 \end{array}$$

3. Subtract. Show your work.

$$\begin{array}{r} 72 \\ - \underline{35} \end{array} \qquad \begin{array}{r} 153 \\ - \underline{\ 28} \end{array}$$

SRB 54 55

4. There are 17 boys and 24 girls in the math club. How many children in all are in the math club?

__________ children

Total	
Part	**Part**

5. About what time is it?

____

6. Jack answered 29 questions. José answered 37 questions. How many fewer questions did Jack answer than José?

__________ questions

Quantity	
Quantity	**Difference**

SRB 190

Date Time

Math Boxes 2.10

1. Which tool would you use to measure the following?

yardstick	ruler	thermometer

temperature thermometer

height of the ceiling yardstick

length of your thumb ruler

2. Circle the best unit of measurement.

distance to Spain
miles centimeters inches

width of a crayon
miles centimeters feet

length of your journal
miles yards inches

3. Measure the line segment in inches.

$2\frac{1}{2}$ inches

4. Measure the line segment in centimeters.

$5\frac{1}{2}$ centimeters

5. How many squares are shaded?

5 squares

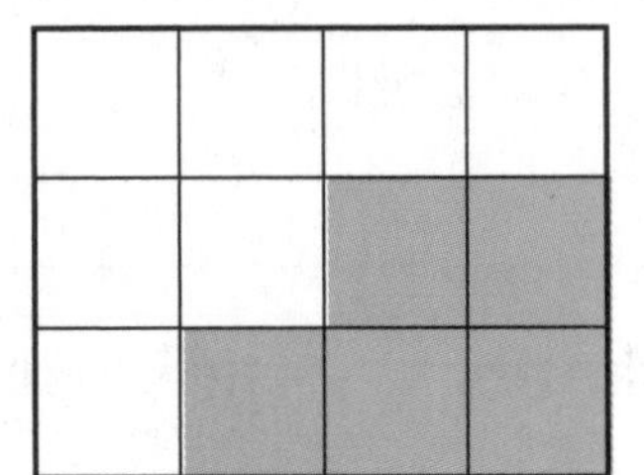

6. How long is the fence around the flowers?

18 feet

3 feet

2 feet 2 feet

3 feet

Date Time

Estimating and Measuring Lengths

Work with a partner. Estimate the lengths of things in the classroom in "class shoe" units. Write the estimate. Then use the "class shoe" strip to measure the object. Write the measurement.

Object	Estimate	Measurement
	about _____ "class shoes"	about _____ "class shoes"
	about _____ "class shoes"	about _____ "class shoes"
	about _____ "class shoes"	about _____ "class shoes"
	about _____ "class shoes"	about _____ "class shoes"
	about _____ "class shoes"	about _____ "class shoes"
	about _____ "class shoes"	about _____ "class shoes"
	about _____ "class shoes"	about _____ "class shoes"
	about _____ "class shoes"	about _____ "class shoes"

Why is it important to use the same units everyone else is using to measure things?

____

____

____

____

Date Time

Addition and Subtraction Practice

Add or subtract. Make a ballpark estimate to check your answer. Write a number model for your estimate.

Unit
pumpkin seeds

1.

$$\begin{array}{r} 681 \\ +\ 253 \\ \hline \end{array}$$

Ballpark estimate:

____

2.

$$\begin{array}{r} 749 \\ +\ 161 \\ \hline \end{array}$$

Ballpark estimate:

____

3.

$$\begin{array}{r} 417 \\ +\ 386 \\ \hline \end{array}$$

Ballpark estimate:

____

4.

$$\begin{array}{r} 472 \\ -\ 253 \\ \hline \end{array}$$

Ballpark estimate:

____

5.

$$\begin{array}{r} 728 \\ -\ 173 \\ \hline \end{array}$$

Ballpark estimate:

____

6.

$$\begin{array}{r} 550 \\ -\ 364 \\ \hline \end{array}$$

Ballpark estimate:

____

Date Time

Math Boxes 3.1

1. Show $10.78 in two other ways.

$5
$5
Ⓠ Ⓠ
Ⓓ Ⓓ
Ⓝ Ⓟ
Ⓟ Ⓟ

2. Find the rule and complete the table.

in	out
117	112
119	
	116
	131
142	

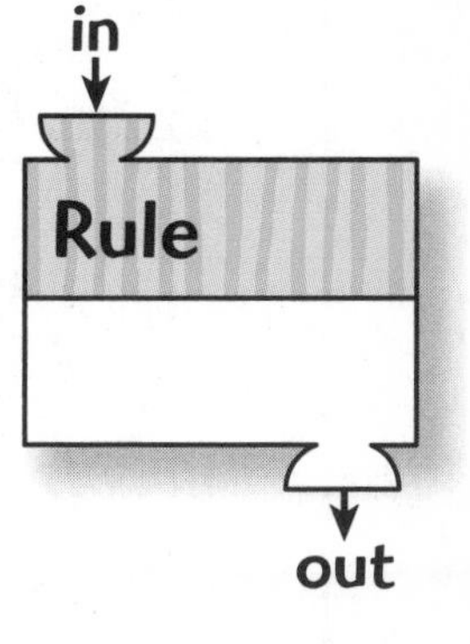

3. Shade to show the following data.
A is 4 cm.
B is 3 cm.
C is 8 cm.
D is 7 cm.

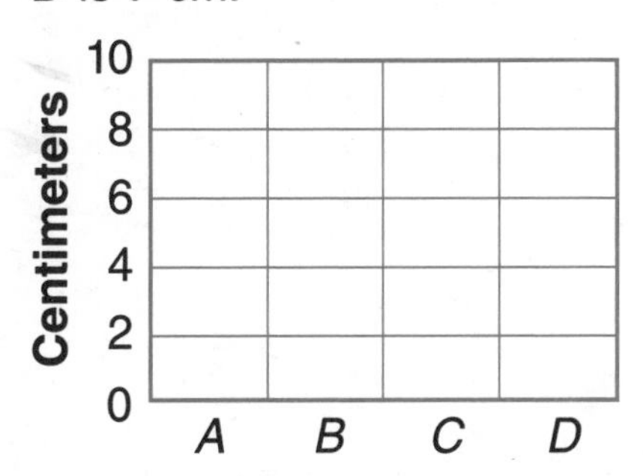

4. Write a number story by filling in the blanks.

Tom collects coins. He has

_____ quarters, _____ dimes,

_____ nickels, and _____ pennies.

How many coins in all?

____

5. Write <, >, or =.

$4 + 5 + 6$ _____ $3 + 5 + 7$

$7 + 5 + 9$ _____ $6 + 6 + 8$

$2 + 11 + 4$ _____ $7 + 1 + 9$

$15 + 7 + 5$ _____ $9 + 9 + 9$

$4 + 5 + 6$ _____ $3 + 7 + 6$

6. Add. Show your work.

$$\begin{array}{r} 492 \\ +\ 18 \\ \hline \end{array} \qquad \begin{array}{r} 555 \\ +\ 192 \\ \hline \end{array}$$

SRB 51 52

Date Time

Measuring Line Segments

1. Use Ruler A to measure to the nearest inch (in.).

   Use Ruler E to measure to the nearest centimeter (cm).

	Ruler A	Ruler E
______________________	about ______ in.	about ______ cm
______________________________________	about ______ in.	about ______ cm
___________________________	about ______ in.	about ______ cm

2. Use Ruler B to measure to the nearest $\frac{1}{2}$ inch.

   Use Ruler E to measure to the nearest $\frac{1}{2}$ centimeter (cm).

	Ruler B	Ruler E
_______________________________	about ______ in.	about ______ cm
_______________	about ______ in.	about ______ cm
__________________________	about ______ in.	about ______ cm

3. Use Ruler C to measure to the nearest $\frac{1}{4}$ inch.

   Use Ruler E to measure to the nearest millimeter (mm).

	Ruler C	Ruler E
_________	about ______ in.	about ______ mm
________________________	about ______ in.	about ______ mm
_________________________________	about ______ in.	about ______ mm

Date Time

Math Boxes 3.2

1. Complete the puzzle.

9,632				
			9,665	

SRB 7 8

2. 53 people were standing in line at 9:00 A.M. 97 people were standing in line at 10:00 A.M. How many more people were standing in line at 10:00 A.M.?

________ people

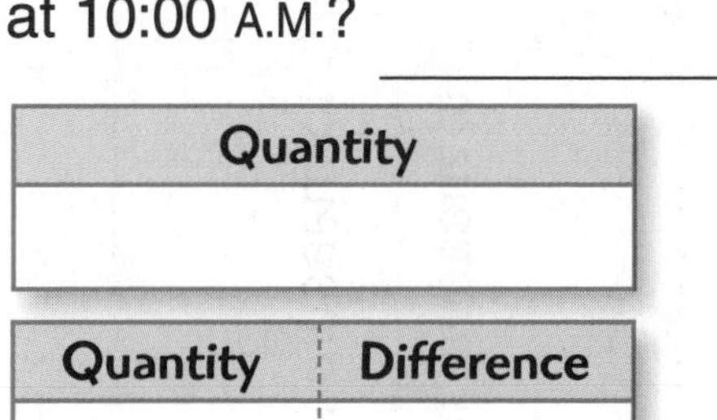

3. Count by 100s.

97; ____; ____;

____; 497; ____;

____; ____; ____;

____; ____; ____

4. Subtract. Show your work.

$$\begin{array}{r} 943 \\ -\ 409 \\ \hline \end{array} \qquad \begin{array}{r} 884 \\ -\ 299 \\ \hline \end{array} \qquad \begin{array}{r} 695 \\ -\ 47 \\ \hline \end{array}$$

5. It is 7:45 A.M. Draw the hour and minute hands to show the time 15 minutes earlier. What time does the clock show now?

____

6. Solve.

____ = 8 + 9

____ = 48 + 9

9 + 5 = ____

900 + 500 = ____

____ = 12 − 4

____ = 12,000 − 4,000

Date Time

Body Measures

Work with a partner to find each measurement to the nearest $\frac{1}{4}$ inch.

	Adult at Home	Me (Now)	Me (Later)
Date	______ , _____	______ , _____	______ , _____
height	about _____ in.	about _____ in.	about _____ in.
shoe length	about _____ in.	about _____ in.	about _____ in.
around neck	about _____ in.	about _____ in.	about _____ in.
around wrist	about _____ in.	about _____ in.	about _____ in.
waist to floor	about _____ in.	about _____ in.	about _____ in.
forearm	about _____ in.	about _____ in.	about _____ in.
hand span	about _____ in.	about _____ in.	about _____ in.
arm span	about _____ in.	about _____ in.	about _____ in.
____________	about _____ in.	about _____ in.	about _____ in.
____________	about _____ in.	about _____ in.	about _____ in.
____________	about _____ in.	about _____ in.	about _____ in.

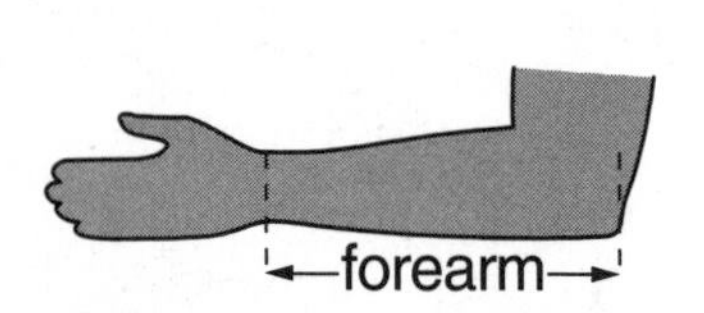

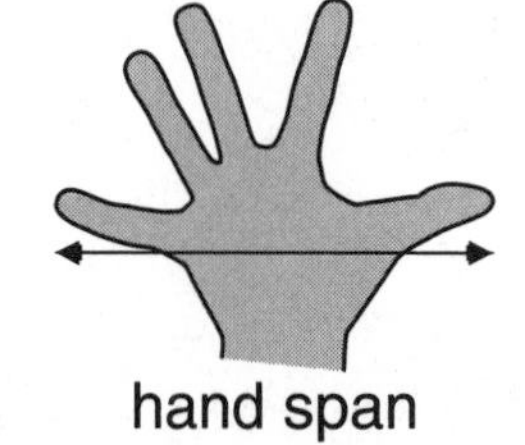

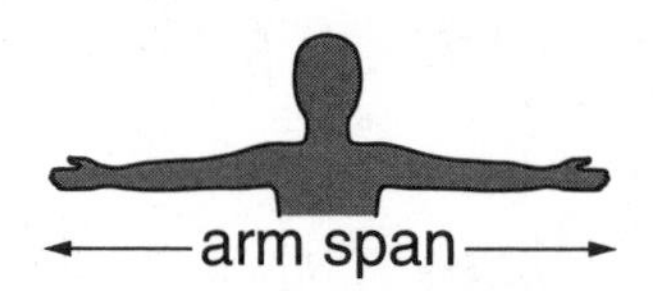

Estimating Lengths

1. Follow these steps using **U.S. customary** units: inches (in.), feet (ft), or yards (yd). Then follow these steps using **metric** units: millimeters (mm), centimeters (cm), decimeters (dm), or meters (m).

   - Use personal references to estimate the measures.
   - Record your estimates. Be sure to write the units.
   - Measure with a ruler or tape measure. Record your measurements.

Objects	U.S. Customary Units		Metric Units	
	Estimate	Measurement	Estimate	Measurement
height of your desk				
long side of your calculator				
short side of the classroom				
distance around your head				

2. Choose your own things to estimate and measure.

Objects	U.S. Customary Units		Metric Units	
	Estimate	Measurement	Estimate	Measurement

Date Time

Math Boxes 3.3

1. Write the number that is

	10 less	100 less	1,000 less
1,067	______	______	______
1,593	______	______	______
2,154	______	______	______
6,163	______	______	______

2. Measure to the nearest $\frac{1}{4}$ inch.

______

Draw a line segment $1\frac{1}{2}$ inches long.

3. Choose a 3-digit number and write at least five names for that number.

4. Fill in the missing amounts.

I had 38¢. I spent ______.
I have 15¢ left.

I had 54¢. I found ______.
Now I have 83¢.

5. 8 + 6 = ____

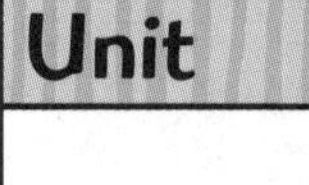

8 + 6 + 7 = ____

8 + 6 + 7 + 5 = ____

$$\begin{array}{r} 17 \\ +\ 8 \\ \hline \end{array} \qquad \begin{array}{r} 17 \\ 8 \\ +\ 5 \\ \hline \end{array} \qquad \begin{array}{r} 17 \\ 8 \\ 5 \\ +\ 19 \\ \hline \end{array}$$

6. Add. Show your work.

$$\begin{array}{r} 384 \\ +\ 675 \\ \hline \end{array} \qquad \begin{array}{r} 8{,}916 \\ +\ 7{,}504 \\ \hline \end{array}$$

Date Time

Perimeters of Polygons

1. Record the **perimeter** (the distance around) of your straw rectangle and parallelogram.

   rectangle: about _____ inches parallelogram: about _____ inches

2. Use a tape measure to find each side and the perimeter.

Polygon	Each Side	Perimeter
triangle	about _____ in., about _____ in., about _____ in.	about _____ in.
triangle	about _____ in., about _____ in., about _____ in.	about _____ in.
square	about _____ in.	about _____ in.
rhombus	about _____ in.	about _____ in.
trapezoid	about _____ in., about _____ in. about _____ in., about _____ in.	about _____ in.

3. Find the perimeter, in inches, of the figures below.

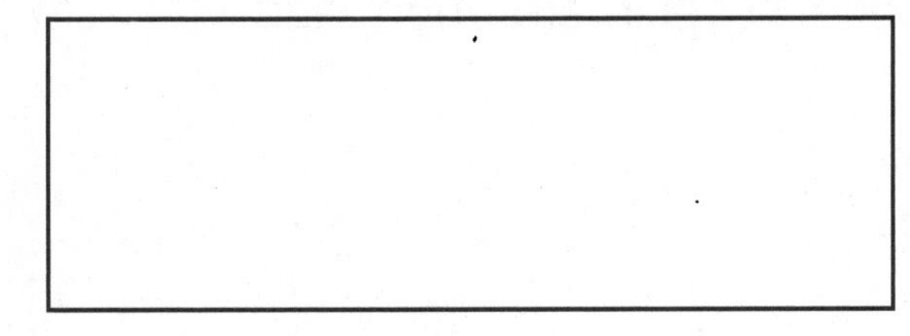

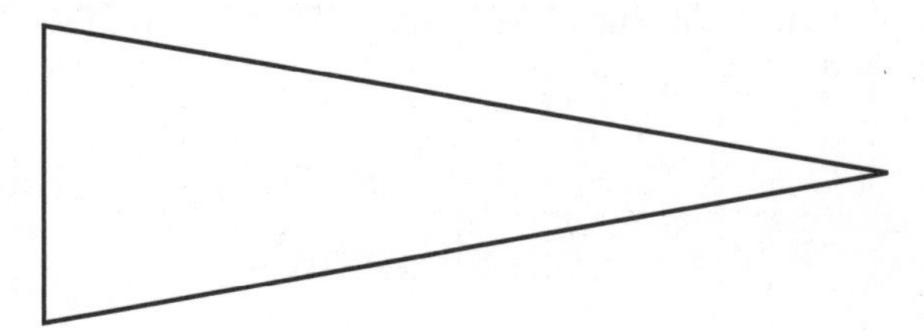

_______________ _______________

4. Draw each shape on the centimeter grid.

   square with perimeter = 16 cm rectangle with perimeter = 20 cm

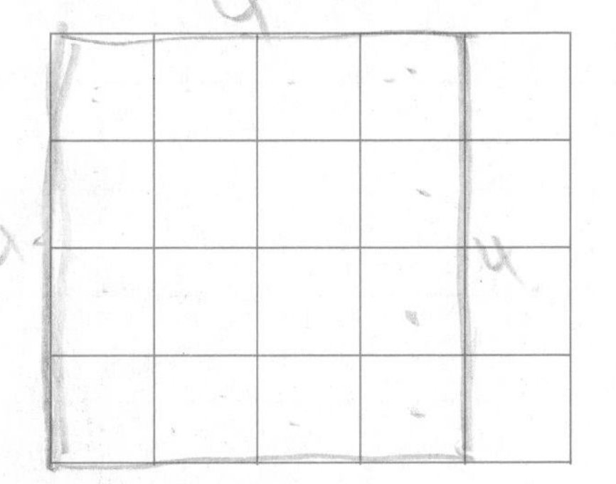

Date Time

Measures Hunt

Find out about how long some objects are.
These objects will be **personal references.**
Use your personal references to estimate the lengths of other things.

1. Find things that are about 1 inch long, 1 foot long, and 1 yard long.
   Use a ruler, tape measure, or yardstick.
   List your objects below.

About 1 inch (in.)	About 1 foot (ft)	About 1 yard (yd)

2. Find things that are about 1 centimeter long, 1 decimeter long, and 1 meter long.
   Use a ruler, tape measure, or meterstick.
   List your objects below.

About 1 centimeter (cm)	About 1 decimeter (dm)	About 1 meter (m)

Math Boxes 3.4

1. "What's My Rule?"

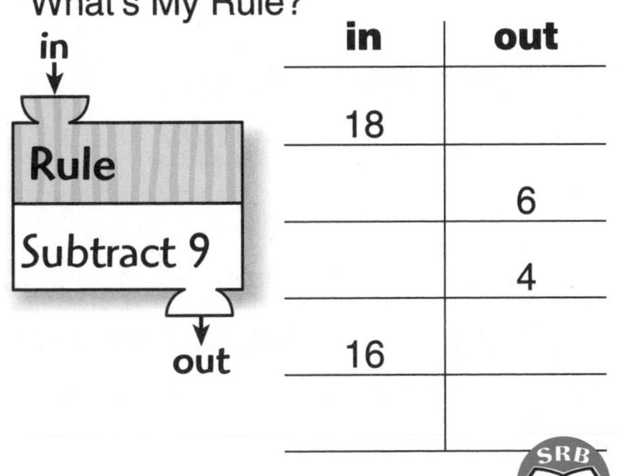

in	out
18	
	6
	4
16	

2. The driving distance between St. Louis and Denver is about 863 miles. If you go by way of Wichita, the distance is about 982 miles. How much farther is it to go by way of Wichita?

______ miles farther

3. Fill in the empty frames and the rule box.

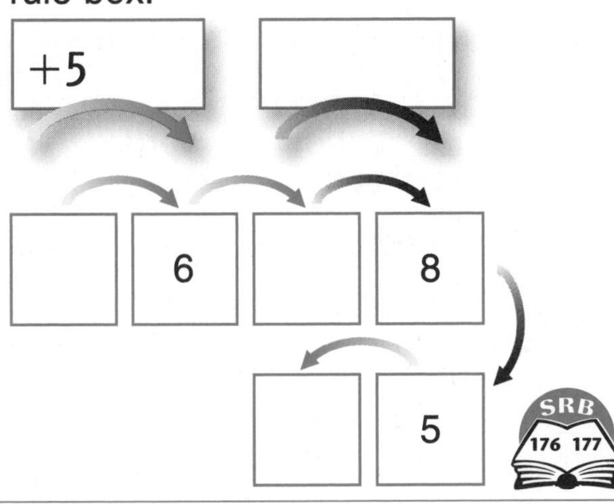

4. Subtract. Show your work.

Unit
buttons

$$\begin{array}{r} 704 \\ -\ 86 \\ \hline \end{array} \qquad \begin{array}{r} 6{,}243 \\ -\ 2{,}948 \\ \hline \end{array}$$

5. Write $<$, $>$, or $=$.

$1\frac{1}{2}$ feet ____ 16 inches

3 feet ____ 2 yards

5 feet ____ 60 inches

55 inches ____ 1 yard

6. Measure to the nearest centimeter.

______

______

Draw a line segment 7 centimeters long.

Date Time

Math Boxes 3.5

1. Find the perimeter.

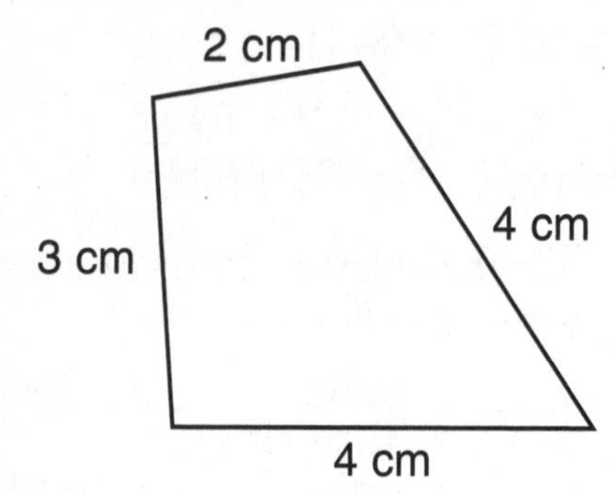

perimeter = 13cm
(unit)

2. Measure to the nearest $\frac{1}{4}$ inch.

______

Draw a line segment $2\frac{1}{2}$ inches long.

3. Yuri saved $24.85. He earned $9.95 more. How much did he have then?

______

$24

4. Write the equivalent lengths.

3 yards = ______ ft

______ inches = 2 yards

50 millimeters = ______ centimeters

3 meters = ______ centimeters

5. Add.

9 + 22 + 11 = ______

13 + 17 + 16 = ______

24 + 6 + 9 = ______

6. Make a ballpark estimate to check that the answer makes sense.

492 + 108 = ______

about ______

______ = 648 + 209

about ______

Date Time

Geoboard Perimeters

Materials ❑ geoboard and rubber bands, or geoboard dot paper

Work with a partner.

1. Suppose that the distance between two pins is 1 unit. Make a rectangle with a perimeter of 14 units. Use rubber bands and a geoboard, or draw the rectangle on dot paper. Record the lengths of the sides in the table.

1 unit

1 unit

2. Now make a different rectangle that also has a perimeter of 14 units. Record the lengths of the sides for this shape.
3. Complete the table for other perimeters.
4. Try to make a rectangle or square with a perimeter of 13 units.
5. Try to make other rectangles or squares with perimeters that are an odd number of units.

Geoboard Perimeters		
Perimeter	**Longer sides**	**Shorter sides**
14 units	_____ units	_____ units
14 units	_____ units	_____ units
14 units	_____ units	_____ units
12 units	_____ units	_____ units
12 units	_____ units	_____ units
12 units	_____ units	_____ units
16 units	_____ units	_____ units
16 units	_____ units	_____ units
16 units	_____ units	_____ units
16 units	_____ units	_____ units

Challenge

Change the unit. Now 1 unit is double the distance between two points. Make a rectangle or square whose perimeter is an odd number of units.

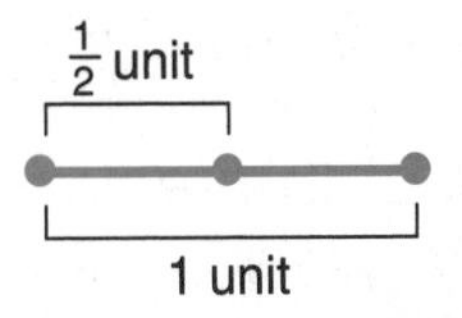

Follow-Up

Look for a pattern in your table. Can you find one? Now, do not use a geoboard or dot paper. Find the lengths of the sides of a rectangle or square with a perimeter of 24 units. Then make or draw the shape to check your answer.

Date   Time

Tiling with Pattern Blocks

Materials ❑ pattern blocks: square, triangle, narrow rhombus

❑ crayons

Work with a partner.

1. Use square pattern blocks. Look at the top rectangle on the next page. Cover as much of the rectangle as you can, placing all of the blocks inside it. There may be uncovered spaces at the edges. Do not overlap the blocks. Line them up so that there are no gaps. This is called **"tiling."**
2. Count and record the number of blocks you used.
3. Trace around the edges of each block. Then color any spaces not covered by blocks. Estimate how many blocks would be needed to cover the colored spaces.
4. Record how many blocks are needed to cover the whole rectangle.
5. Tile the second rectangle with triangles. Repeat Steps 2–4 above.
6. Tile the third rectangle with narrow rhombuses. Repeat Steps 2–4 above.

Follow-Up

7. The **area** of a shape is a measure of the space inside the shape. You measured the area of a rectangle three ways: with squares, triangles, and narrow rhombuses. Record the areas below.

   The area of the rectangle is about _____ squares.

   The area of the rectangle is about _____ triangles.

   The area of the rectangle is about _____ narrow rhombuses.

8. Which of the three pattern blocks has the largest area? ____________________

   Which has the smallest area? ____________________

   How did you decide? ______________________________

   ____

Date Time

Tiling with Pattern Blocks (cont.)

Cover this rectangle with squares.

About _____ squares cover the whole rectangle.

Cover this rectangle with triangles.

About _____ triangles cover the whole rectangle.

Cover this rectangle with narrow rhombuses.

About _____ narrow rhombuses cover the whole rectangle.

Date Time

Straw Triangles

Materials ❑ 4-inch, 6-inch, and 8-inch straws

❑ twist-ties

Work in a group to make as many different-size triangles as you can out of the straws and twist-ties. (Be sure that straws are touching at all ends.) Before you start, decide how you will share the work.

For each triangle, record the length of each side and the perimeter in the chart. The triangle made out of the shortest straws is already recorded.

Straw Triangles			
Side 1	Side 2	Side 3	Perimeter
4 in.	4 in.	4 in.	12 in.
______	______	______	______
______	______	______	______
______	______	______	______
______	______	______	______
______	______	______	______
______	______	______	______
______	______	______	______
______	______	______	______

Follow-Up

Discuss these questions with others in your group.

1. Which triangles have similar shapes?

____

2. Which pairs of triangles have the same perimeter?

____

3. By looking at your constructions, estimate which triangle of each pair of triangles in problem 2 has the larger area (space inside the triangles).

____

4. What happens if you try to make a triangle out of two 4-inch straws and one 8-inch straw?

____

Date Time

Areas of Rectangles

Draw each rectangle on the grid. Make a dot inside each small square in your rectangle.

1. Draw a 3-by-5 rectangle.

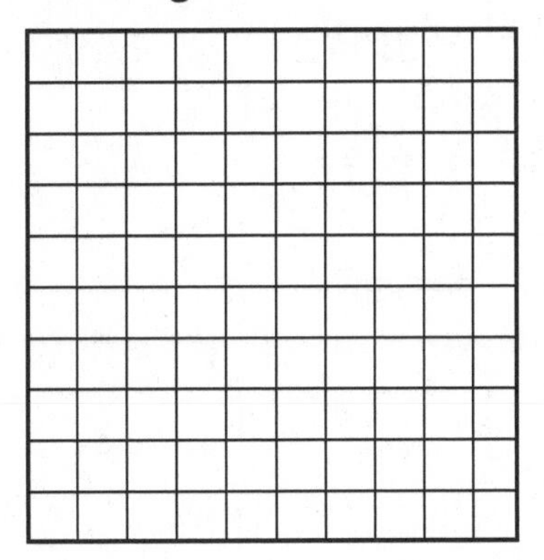

Area = _____ square units

2. Draw a 6-by-8 rectangle.

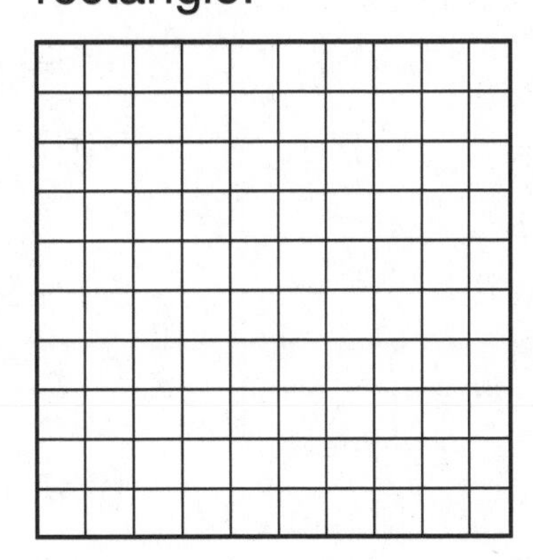

Area = _____ square units

3. Draw a 9-by-5 rectangle.

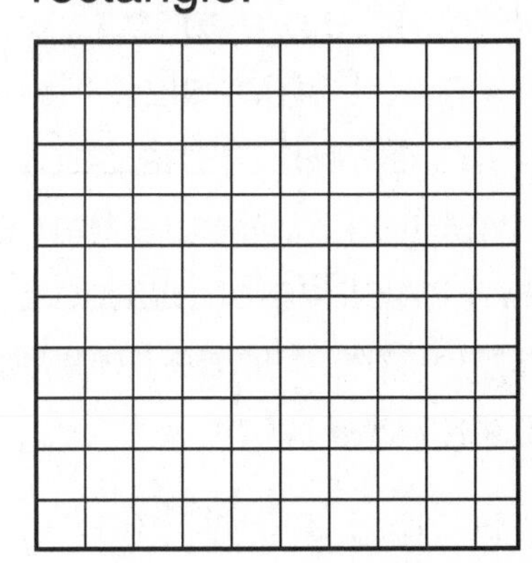

Area = _____ square units

Fill in the blanks.

4.

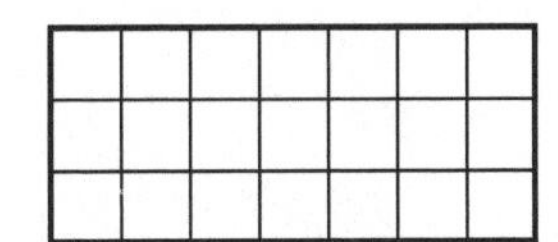

This is a _____-by-_____ rectangle.

Area = _____ square units

5.

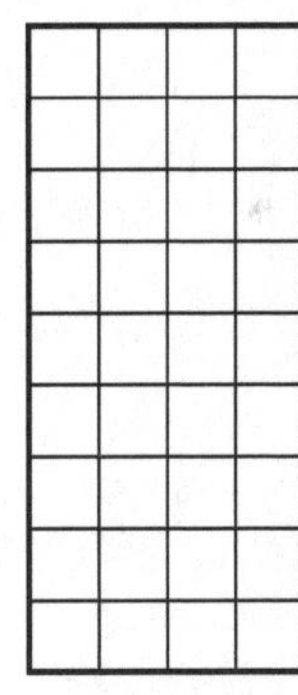

This is a _____-by-_____ rectangle.

Area = _____ square units

6.

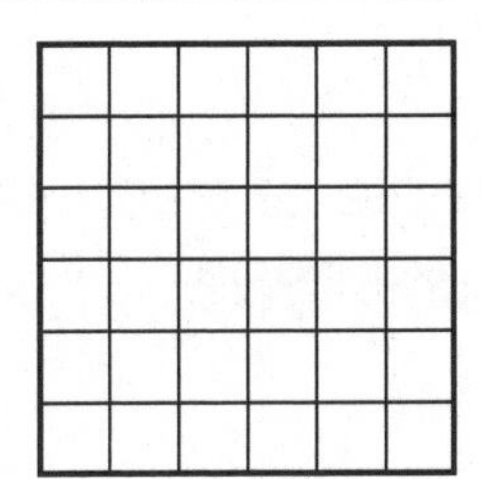

This is a _____-by-_____ rectangle.

Area = _____ square units

7.

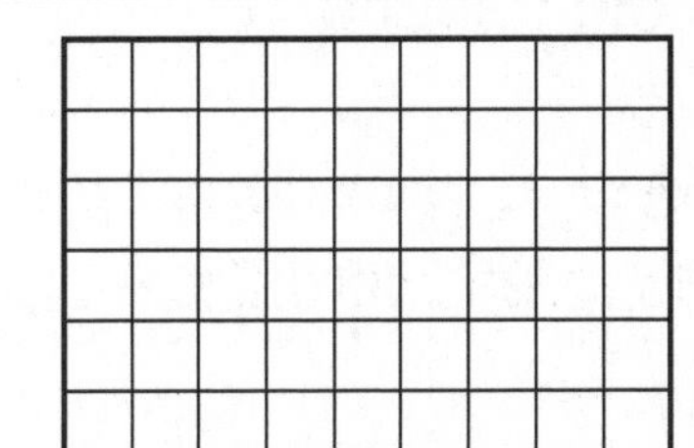

This is a _____-by-_____ rectangle.

Area = _____ square units

Date Time

Math Boxes 3.6

1. Put these numbers in order from smallest to largest:

47,912 ______ ← smallest

47,192 ______

49,271 ______

49,172 ______ ← largest

2. Solve.

Unit

______ = 7 + 9

______ = 37 + 9

16 − 8 = ______

76 − 8 = ______

6 + 5 = ______

600 + 500 = ______

3. There were 144 cartons of milk delivered to school. 84 of the cartons were chocolate milk. The rest were 2% milk. How many cartons of 2% milk were delivered?

______ cartons

4. Subtract. Show your work.

$$\begin{array}{r} 384 \\ -\ 175 \\ \hline \end{array} \qquad \begin{array}{r} 8{,}306 \\ -\ 7{,}574 \\ \hline \end{array}$$

5. When I left home, I had $4.00. I spent 73¢ at the fruit stand and $2.59 at the grocery store. How much did I spend in all?

______

How much do I have when I go home?

______

6. Measure to the nearest centimeter.

______

Draw a line segment 4 centimeters long.

Date Time

More Areas of Rectangles

Make a dot inside each small square in one row. Then fill in the blanks.

1.

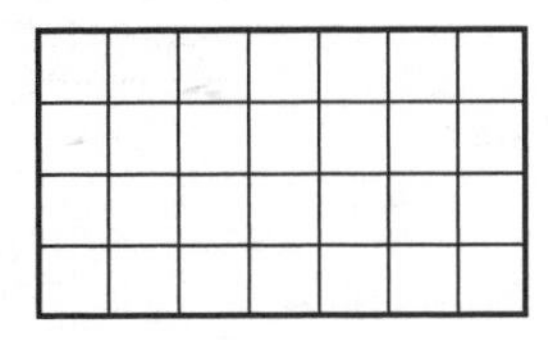

Squares in a row: ____

Number of rows: ____

Number model:

____ × ____ = ____

Area = ____ square units

2.

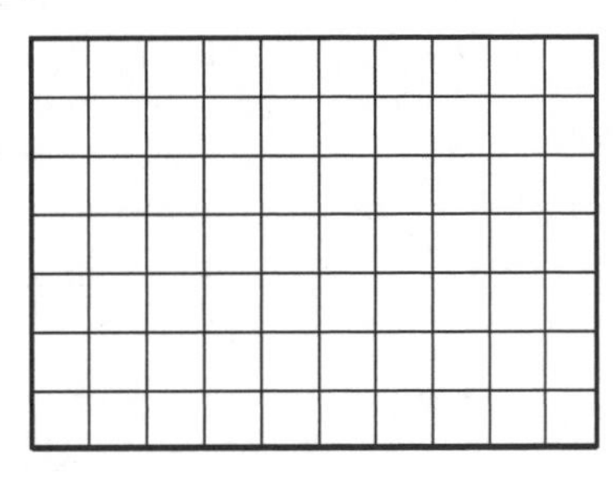

Squares in a row: ____

Number of rows: ____

Number model:

____ × ____ = ____

Area = ____ square units

3.

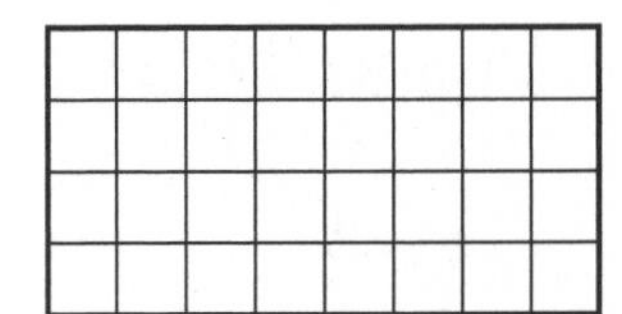

Squares in a row: ____

Number of rows: ____

Number model:

____ × ____ = ____

Area = ____ square units

Now, draw the rectangle on the grid. Then fill in the blanks.

4. Draw a 5-by-7 rectangle.

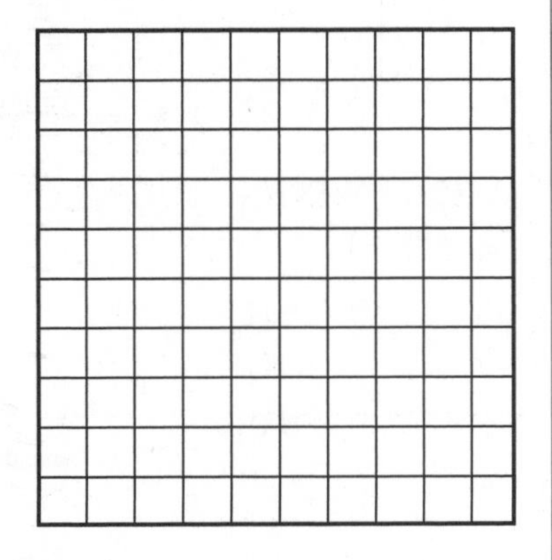

Number model:

____ × ____ = ____

Area = ____ square units

5. Draw an 8-by-8 rectangle.

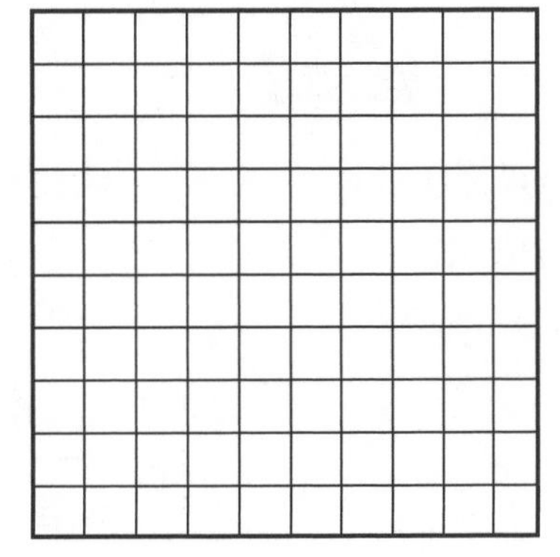

Number model:

____ × ____ = ____

Area = ____ square units

6. Draw a 3-by-9 rectangle.

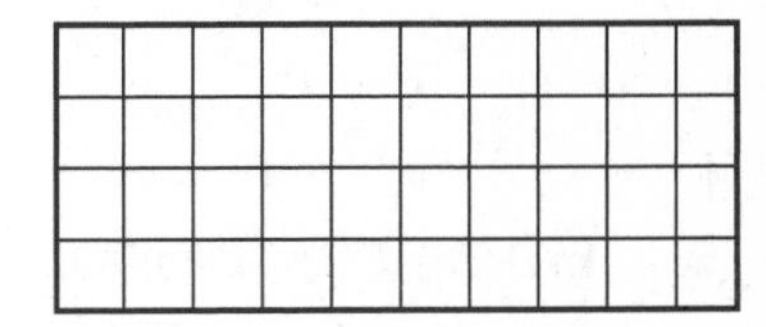

Number model:

____ × ____ = ____

Area = ____ square units

Date Time

Math Boxes 3.7

1. What is the perimeter?

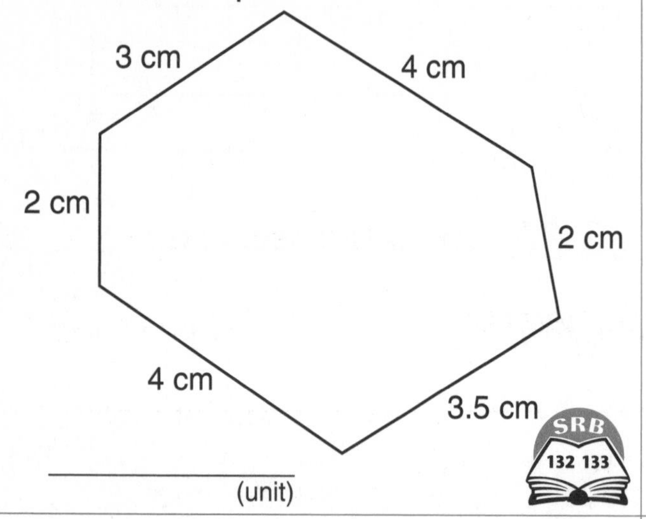

_______________ (unit)

SRB 132 133

2. Measure to the nearest $\frac{1}{4}$ inch.

____

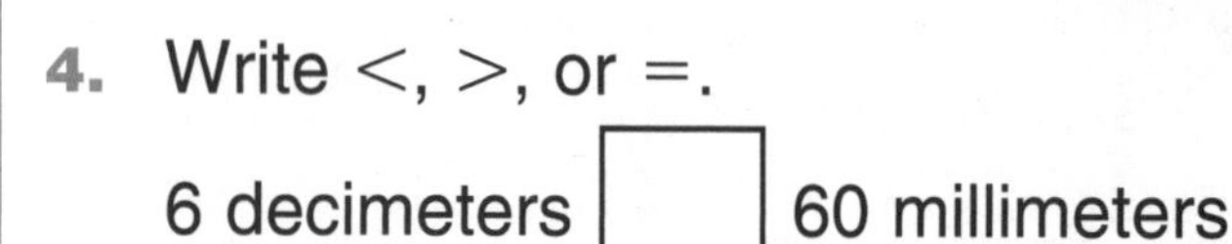

Draw a line segment $2\frac{3}{4}$ inches long.

3. At 7:00 A.M., the temperature was 23°F. At 10:00 A.M., the temperature was 40°F. How much warmer was it at 10:00 A.M. than at 7:00 A.M.?

__________ °F warmer

4. Write $<$, $>$, or $=$.

6 decimeters ☐ 60 millimeters

3 yards ☐ 36 inches

2 centimeters ☐ 4 meters

Write your own.

5. Complete the number story.

Amber ate _____ grapes.

Zack ate _____ grapes.

Sophie ate _____ grapes.

_____ grapes were eaten in all.

6. Add. Show your work.

$$\begin{array}{r} 38 \\ 698 \\ +\ 202 \\ \hline \end{array} \qquad \begin{array}{r} 182 \\ 309 \\ 962 \\ +\ 745 \\ \hline \end{array}$$

Date Time

Diameters and Circumferences

1. Find numbers on the label of your can. Write some of them below. Also write the unit if there is one.

_______________ _______________

2. Record the diameter and circumference of your can.

can letter _____ **diameter:** about _____ cm **circumference:** about _____ cm

3. Write the rule linking diameter and circumference:

____

Review

4.

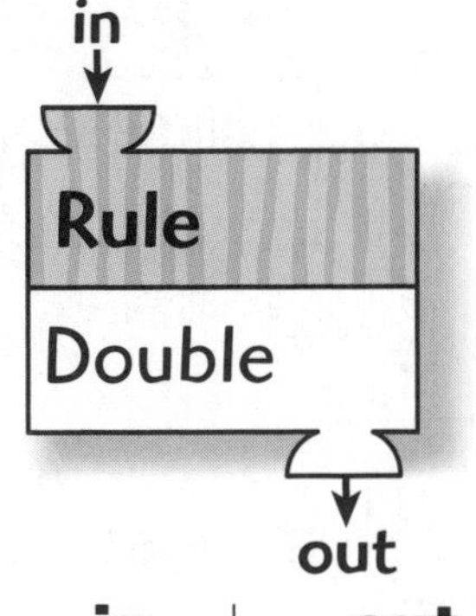

in	out
5	
50	
500	
5,000	

5.

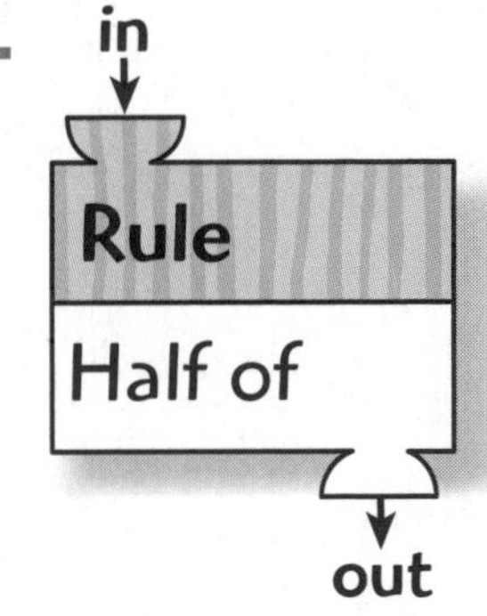

in	out
12	
120	
1,200	
12,000	

6.

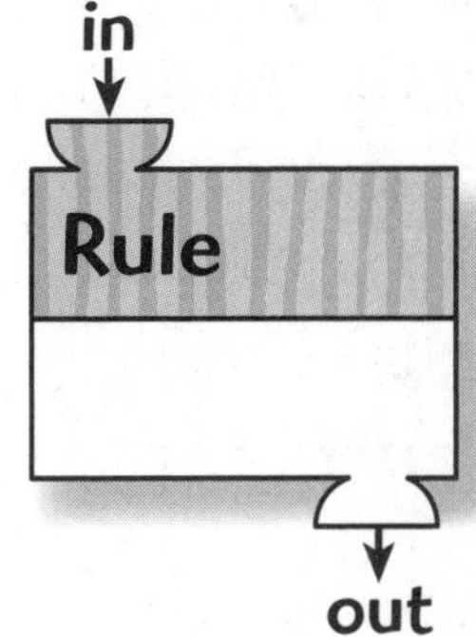

in	out
3	6
	20
5	10
70	
	400

7.

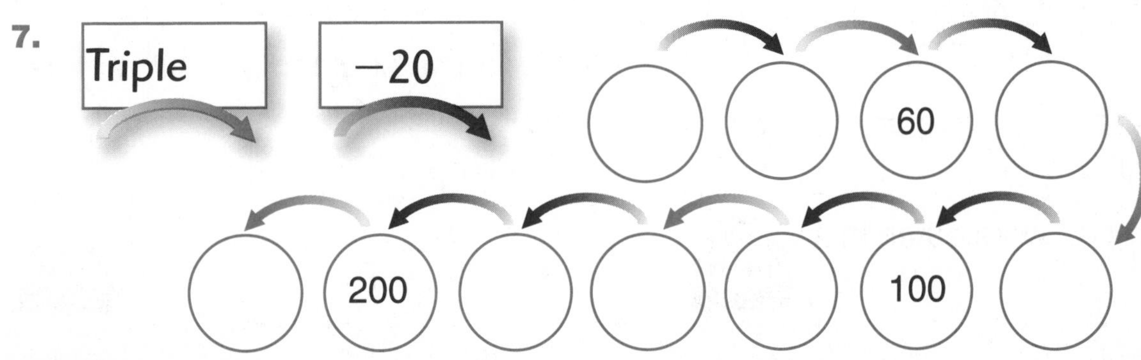

Date Time

Math Boxes 3.8

1.

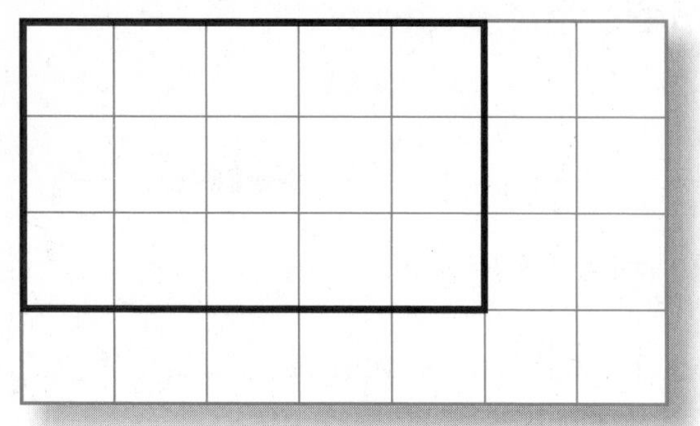

Area: _____ square cm

2. Subtract.

49 − 17 = ________

69 − 17 = ________

199 − 17 = ________

2,119 − 17 = ________

9,139 − 17 = ________

3. Find the total value.

4 [$1]

3Ⓠ

6Ⓓ

2Ⓝ

7Ⓟ

Total $________

4. Subtract. Show your work.

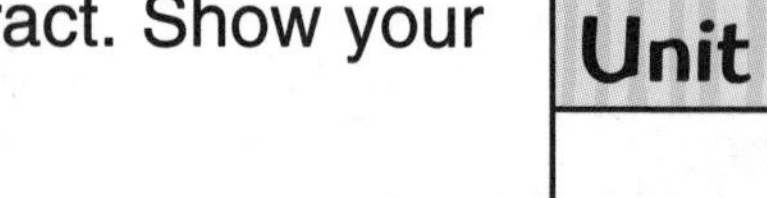

$$\begin{array}{r} 563 \\ -\ 294 \\ \hline \end{array} \qquad \begin{array}{r} 807 \\ -\ 429 \\ \hline \end{array}$$

5.

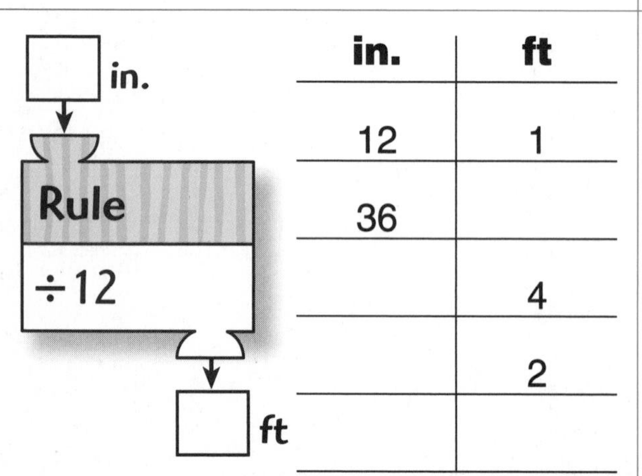

in.	ft
12	1
36	
	4
	2

6. Measure to the nearest millimeter.

Draw a line segment 20 millimeters long.

Use with Lesson 3.8.

Date Time

Units of Linear Measure

Choose a U.S. customary unit and a metric unit for each object.
Put a check in the box.

	U.S. Customary				Metric			
	in.	ft	yd	mi	mm	cm	m	km
thickness of a magazine								
length of hair								
diameter of a quarter								
height of a building								
distance to Paris								
length of a baseball bat								
circumference of a telephone pole								
perimeter of a baseball diamond								
depth of a lake								
Make up your own.								

Math Boxes 3.9

1. How many rows? _____

   How many columns? _____

   How many dots in all? _____

2. There are 3 cars. 4 people are riding in each car. How many people in all?

   _______ people

3. 2 children share 12 toys equally. How many toys does each child get?

   _____ toys

4. Each child has 4 lollipops. There are 16 lollipops. How many children are there?

   _____ children

5. Three children share 10 sticks of gum equally. How many sticks does each child get?

   _____ stick(s)

   How many sticks are left over?

   _____ stick(s)

6. $5 \times 0 =$ _____

   $1 \times 8 =$ _____

   $2 \times 3 =$ _____

   _____ $= 5 \times 3$

   _____ $= 4 \times 10$

Date Time

Solving Multiplication Number Stories

Use the Variety Store Poster on page 239 of the *Student Reference Book.*

For each number story:

- Fill in a multiplication/division diagram with the numbers you know. Write ? for the number you need to find.
- Use counters, draw pictures, or do whatever helps you find the answer.
- Record the answer with its unit. Check whether your answer makes sense.

1. Yosh has 4 boxes of mini stock cars. How many cars does he have?

Answer: 16 cars (unit)

boxes	cars per box	total number of cars
4	4	16

2. How many cards are in 5 packages of file cards?

Answer: 25 packages (unit)

packages	cards per package	total number of cards
5	5	25

3. Claire buys 8 packages of fashion pens. How many pens does she have?

Answer: 64 fashion pens (unit)

packages	pens per package	total number of pens
8	8	64

4. If your mother buys 2 packages of bright shoelaces, how many pairs of shoelaces does she buy?

Answer: 4 (unit)

packages	pairs of shoelaces per package	total number of pairs of shoelaces
2	2	4

Bonus: About how much do the 2 packages cost? 4

Date Time

Writing Multiplication Number Stories

Write 2 multiplication stories. For each story:

- Fill in the multiplication/division diagram. Write ? for the number you need to find.
- Use counters, draw pictures, or do whatever helps you find the answer.
- Record your answer with its unit. Check whether your answer makes sense.

1. ______________________________

____

____

Answer: ______________ (unit)

2. ______________________________

____

____

Answer: ______________ (unit)

Use with Lesson 4.1.

Date Time

Measuring Line Segments

Use your ruler to measure each line segment.

Measure to the nearest $\frac{1}{2}$ inch.

1. ______________________________

about _____ inches

2. ___________________

about _____ inches

3. ____

about _____ inches

Measure to the nearest $\frac{1}{4}$ inch.

4. ________________________

about _____ inches

5. ___________________________________

about _____ inches

Measure as precisely as you can.

6. ______________________________________

about _____ inches

Date Time

Math Boxes 4.1

1. Find the perimeter.

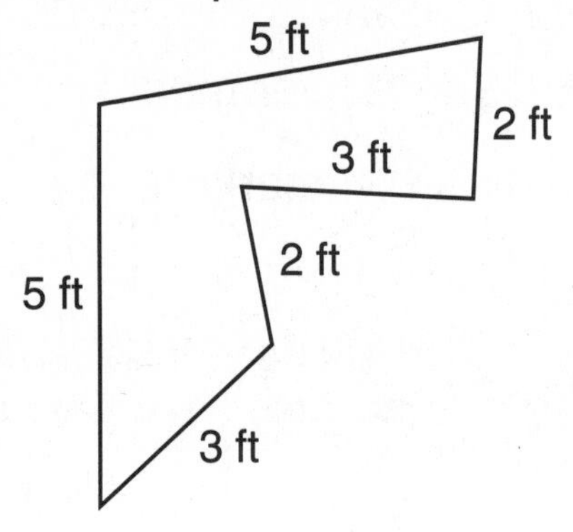

Perimeter = ________ (unit)

2. Measure to the nearest $\frac{1}{4}$ inch.

____

Draw a line segment $1\frac{1}{4}$ inches long.

3. Solve.

12,469 + 10 = ________

12,469 + 100 = ________

12,469 + 1,000 = ________

12,469 + 10,000 = ________

4. Write <, >, or = .

3 decimeters _____ 30 millimeters

$1\frac{1}{2}$ yards _____ 24 inches

45 centimeters _____ 1 meter

9 feet _____ 3 yards

5. Circle the names that belong in the box.

100 − 44	93 − 27	33 + 13
86 − 30	8 × 7	26 + 30
46 + 15	20 + 20 + 16	

6. What is the total value of the coins?

6 Ⓠ
4 Ⓓ
3 Ⓝ
2 Ⓟ

Total value: $________

Date Time

More Multiplication Number Stories

- Fill in the multiplication/division diagram.
- Make an array with counters. Mark the dots to show the array.
- Find the answer. Write the unit with your answer. Write a number model.

1. Mrs. Kwan has 3 boxes of scented markers. Each box has 8 markers. How many markers does she have?

boxes	markers per box	total number of markers

Answer: ______________ (unit) Number model: ______________

2. Monica keeps her doll collection in a case with 5 shelves. On each shelf there are 6 dolls. How many dolls are in Monica's collection?

shelves	dolls per shelf	total number of dolls

Answer: ______________ (unit) Number model: ______________

3. During the summer Jack mows lawns. He can mow 4 lawns per day. How many lawns can he mow in 7 days?

days	lawns per day	total number of lawns

Answer: ______________ (unit) Number model: ______________

Date Time

Perimeter

Measure the perimeter in inches of each figure.

1.

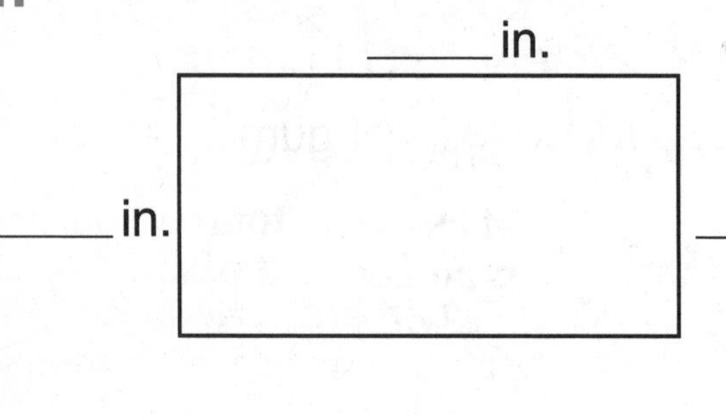

Perimeter: _____ inches

2.

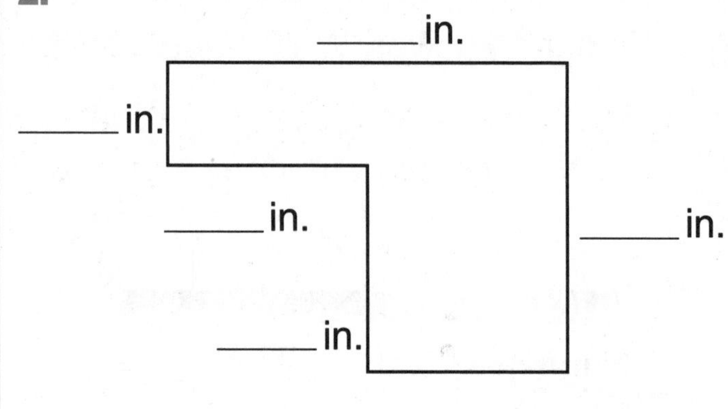

Perimeter: _____ inches

3.

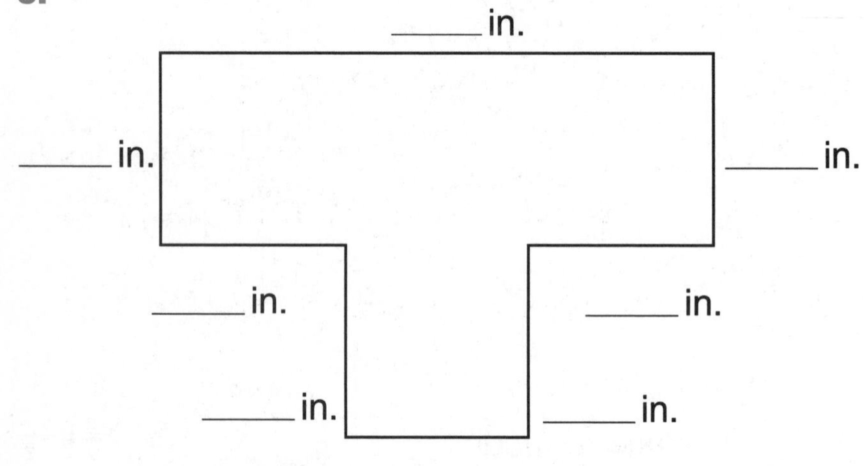

Perimeter: _____ inches

4.

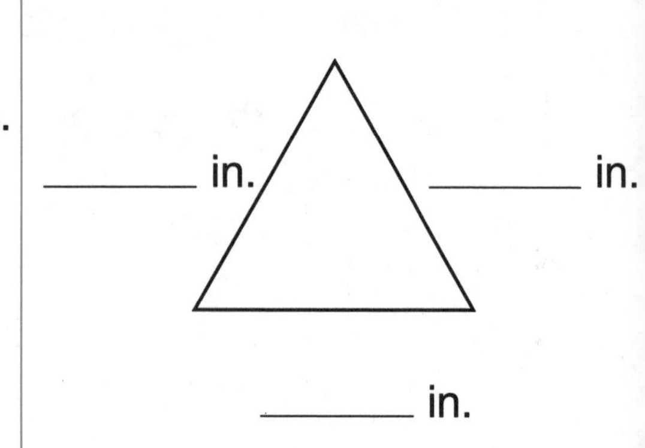

Perimeter: _____ inches

5. Draw any figure with a perimeter of 20 centimeters.

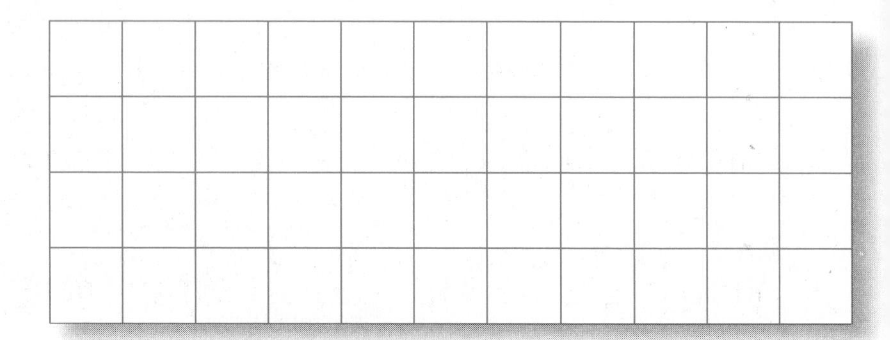

Date Time

Math Boxes 4.2

1. Draw a 2 × 4 rectangle.

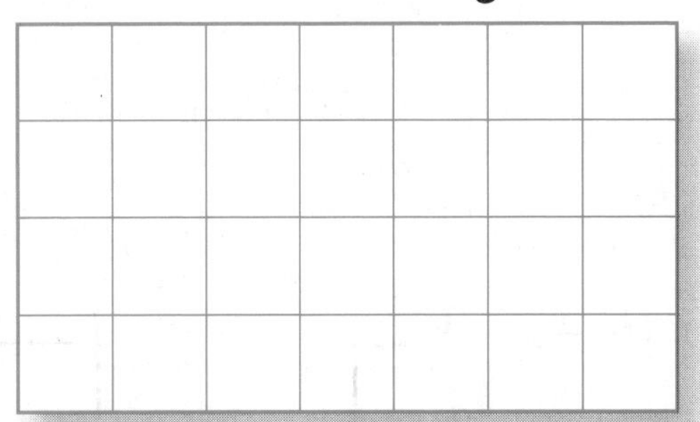

Number model: ____ × ____ = ____

Area: _____ square units

2. 10 packs of gum on the shelf in the candy store. 8 sticks of gum per pack.
How many sticks of gum in all?

packs	sticks of gum per pack	total number of sticks of gum

3. Fill in the numbers.

1,002; 1,001; 1,000; _______;

_______; _______

14,116; 14,117; 14,118; _______;

_______; _______

5,097; 5,098; _______; _______;

_______; _______

4. Fill in the number grid.

2,946

5. Put these units of measure in order from smallest to largest.

mile _______

foot _______

yard _______

inch _______

6. Measure to the nearest centimeter.

____

Draw a line segment 5 centimeters long.

Date Time

Division Practice

Use counters to find the answers. Fill in the blanks.

16 cents shared equally

1. by 2 people:

   _____ ¢ per person

   _____ ¢ remaining

2. by 3 people:

   _____ ¢ per person

   _____ ¢ remaining

3. by 4 people:

   _____ ¢ per person

   _____ ¢ remaining

25¢ shared equally

4. How many people get 5¢?

   _____ people

   _____ ¢ remaining

5. How many people get 3¢?

   _____ people

   _____ ¢ remaining

6. How many people get 6¢?

   _____ people

   _____ ¢ remaining

30 stamps shared equally

7. by 10 people:

   _____ stamps per person

   _____ stamps remaining

8. by 5 people:

   _____ stamps per person

   _____ stamps remaining

9. by 6 people:

   _____ stamps per person

   _____ stamps remaining

10. 21 days
    7 days per week

    _____ weeks

    _____ days remaining

11. 32 crayons
    6 crayons per box

    _____ boxes of crayons

    _____ crayons remaining

12. 24 eggs
    6 eggs per row

    _____ rows of eggs

    _____ eggs remaining

13. There are 18 counters in an array. There are 6 rows.

    How many counters are in each row? _____ counters per row

14. Five children share 12 markers equally. How many markers does each child get? _____ markers _____ markers remaining

Date Time

Math Boxes 4.3

1. Find the perimeter.

Unit
cm

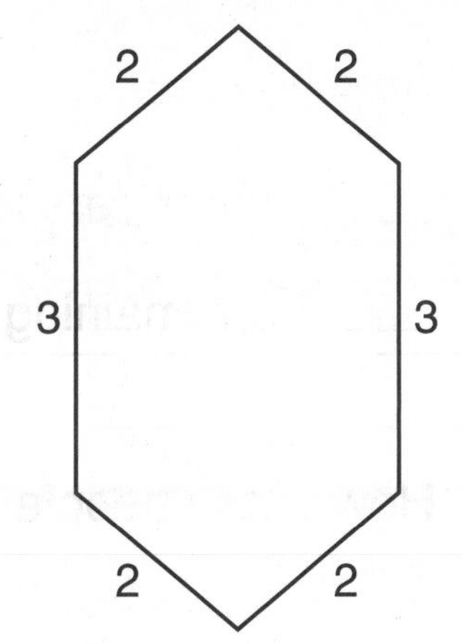

_______ cm

This shape is a

_______________.

2. Make a 4-by-4 array. Complete the number model.

. . . . . .
. . . . . .
. . . . . .
. . . . . .
. . . . . .
. . . . . .

_______ × _______ = _______

SRB 63 64

3. Solve.

45,582 − 10 = _______________

45,582 + 100 = _______________

45,582 + 1,000 = _______________

45,582 − 10,000 = _______________

SRB 18 19

4. Put these metric units of measure in order from smallest to largest.

centimeter _______________

kilometer _______________

millimeter _______________

meter _______________

5. Draw the hands to show 10:20.

How many minutes

until 11:10? _______________

6. Complete.

yd ↓

Rule
×3

↓ ft

yd	ft
2	
5	
	9
	30

Date Time

Solving Multiplication and Division Number Stories

Solve each number story. Use counters, draw an array, or do whatever helps you find the answer. Fill in the diagrams and write number models.

1. Robert has 3 packages of pencils. There are 12 pencils in each package. How many pencils does Robert have in all?

Answer: ______________ (unit)

Number model: ______________

packages	pencils per package	total number of pencils

2. Robert gives 3 pencils to each of his friends. How many friends will get 3 pencils each?

Answer: ______________ (unit)

Number model: ______________

friends	pencils per friend	total number of pencils

3. What if Robert shares his pencils equally among himself and 11 friends? How many pencils does each child get?

Answer: ______________ (unit)

Number model: ______________

Robert and friends	pencils per friend	total number of pencils

4. A class of 30 children wants to play ball. How many teams can be made with exactly 6 children on each team?

Answer: ______________ (unit)

Number model: ______________

teams	children per team	total number of children

5. The same class of 30 children wants to have exactly 4 children on each team. How many teams can be made?

Answer: ______________ (unit)

Number model: ______________

teams	children per team	total number of children

Math Boxes 4.4

1. Draw a shape with an area of 9 square centimeters.

2. Draw an array and complete a number model to match the diagram.

packs	cards per pack	total number of cards
3	6	?

Number model: ______________

3. Fill in the rule and the empty frames.

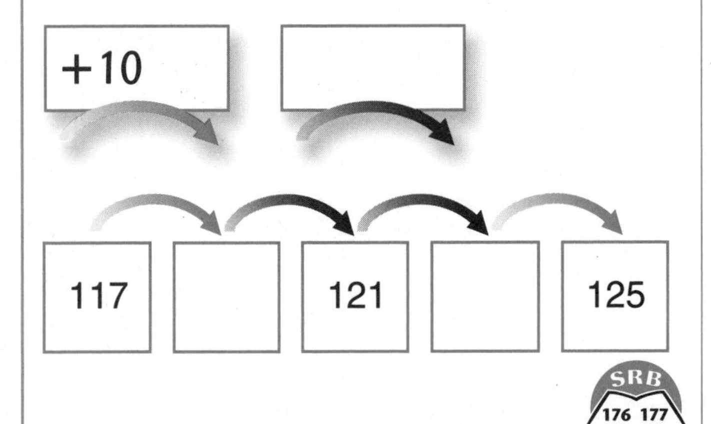

SRB 176 177

4. Scientists studying green turtles counted 136 eggs in a clutch of eggs. 87 eggs did not hatch.

How many eggs did hatch?

_____ eggs

5. Add. Show your work.

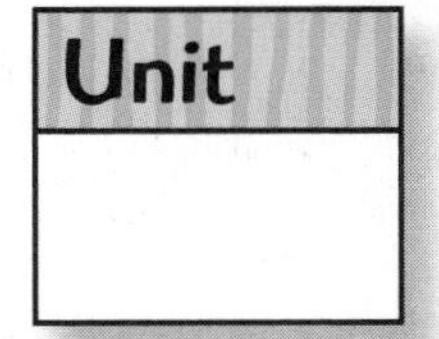

$$\begin{array}{r} 478 \\ +\ 236 \\ \hline \end{array} \qquad \begin{array}{r} 309 \\ +\ 2{,}047 \\ \hline \end{array}$$

6. Write $<$, $>$, or $=$.

$3 + 8 + 7$ _____ $4 + 8 + 6$

$7 + 7 + 9$ _____ $9 + 9 + 5$

$9 + 1 + 8$ _____ $11 + 5 + 3$

$8 + 8 + 8$ _____ $15 + 5 + 7$

$5 + 35 + 17$ _____ $15 + 18 + 25$

SRB 13

Date Time

Math Boxes 4.5

1. Use counters to solve.

Some children are sharing 22 marbles equally. Each child gets 6 marbles.

How many children are sharing?

______________ (unit)

How many marbles are left over?

______________ (unit)

2. Draw Xs in a 5-by-9 array.

How many Xs? _____

Write a number model for the array.

____

3. Subtract. Show your work.

$$\begin{array}{r} 406 \\ -\ 46 \\ \hline \end{array} \qquad \begin{array}{r} 5{,}168 \\ -\ 2{,}936 \\ \hline \end{array}$$

4. Add.

_____ = 47 + 192

_____ = 147 + 292

_____ = 247 + 392

5. Fill in the unit box. Write the missing number in the diagram.

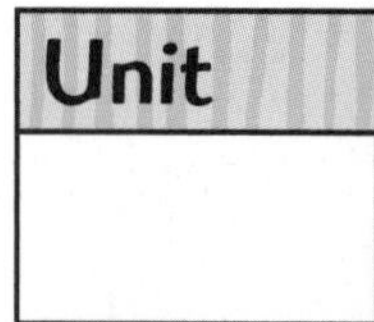

Start	Change	End
	 +107	392

Write a number model.

_____ + _____ = _____

6. Measure to the nearest $\frac{1}{4}$ inch.

____

____

Draw a line segment 3 inches long.

SRB 125 126

Date Time

Math Boxes 4.6

1. On the centimeter grid below, draw a shape with an area of 12 square centimeters.

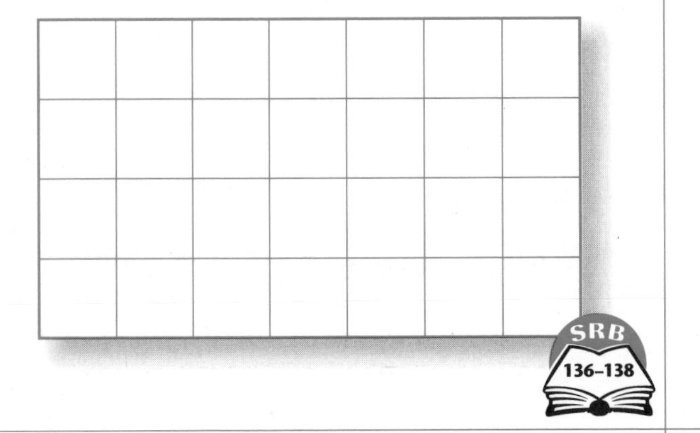

SRB 136–138

2. Write a multiplication story by filling in the blanks.

8 rows of ______________

6 ______________ in each row

How many ______________ in all?

Write a number model.

____

3. Solve.

$2 \times 9 =$ ____

$4 \times 0 =$ ____

____ $= 66 \times 1$

$7 \times$ ____ $= 70$

____ $= 5 \times 8$

4. Write the number sixteen thousand, three hundred two.

____

Write the words for 12,015.

____

____

5. Justin bought 2 gallons of milk. Each gallon cost $2.79. He paid with a $10 bill. How much change did he receive?

____

6. Find the perimeter of the trapezoid.

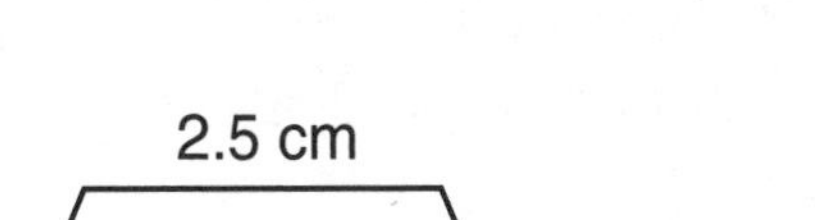

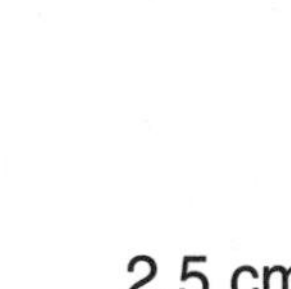

2.5 cm

2.5 cm

2.5 cm

4.5 cm

Perimeter: ________ (unit)

SRB 132 133

Date Time

Math Boxes 4.7

1. Make equal groups.

30 days

make _____ weeks

with _____ days left over.

56 pennies

make _____ quarters

with _____ pennies left over.

2. Draw a 6-by-3 array.

What is the number model?

_____ × _____ = _____

3. Solve.

2 × 2 = _____

16 = 4 × _____

3 × 4 = _____

5 × 6 = _____

18 = _____ × 3

7 × 4 = _____

4. Write the ×, ÷ fact family for the numbers 3, 8, and 24.

24 = _____ × _____

24 = _____ × _____

_____ = 24 ÷ _____

_____ = _____ ÷ _____

5. Add. Show your work.

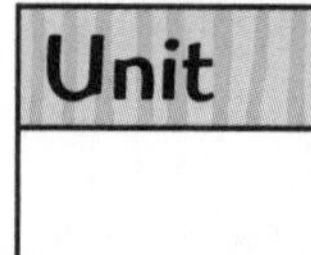


```
  881        6,709
+ 746      +   448
```

6. There are 46 trees and 25 flowers. How many more trees are there than flowers?

_____ trees

Write a number model.

____

Date Time

How Many Dots?

Materials ❑ square pattern blocks

❑ calculator

1. Estimate how many dots are in the array at the right.

   About _______ dots

Make another estimate.
Follow these steps:

2. Cover part of the array with a square pattern block. About how many dots can you cover with one block?

   _________ dots

3. Cover the array. Use as many square pattern blocks as you can. Do not go over the borders of the array. How many blocks did you use?

   _________ blocks

4. Use the information in Steps 2 and 3 to estimate the total number of dots in the array. About _________ dots

Challenge

5. Try to find the exact number of dots in the array.
   Use a calculator to help you. Total number of dots = _________.

Follow-Up

Describe how you found the exact number of dots. _________________________

____

____

Setting up Chairs

1. Record the answer to the problem about setting up chairs from *Math Masters,* page 52.

   There were __________ chairs in the room.

2. Circle dots below to show how the chairs were set up for each of the clues.

Rows of 2	Rows of 3	Rows of 4	Rows of 5
• •	• • •	• • • •	• • • • •
• •	• • •	• • • •	• • • • •
• •	• • •	• • • •	• • • • •
• •	• • •	• • • •	• • • • •
• •	• • •	• • • •	• • • • •
• •	• • •	• • • •	• • • • •
• •	• • •	• • • •	• • • • •
• •	• • •	• • • •	• • • • •
• •	• • •	• • • •	• • • • •
• •	• • •	• • • •	• • • • •
• •	• • •	• • • •	• • • • •
• •	• • •	• • • •	• • • • •
• •	• • •	• • • •	• • • • •
• •	• • •	• • • •	• • • • •
• •	• • •	• • • •	• • • • •
• •	• • •	• • • •	• • • • •
• 1 left over	• 1 left over	• 1 left over	0 left over

Date Time

Math Boxes 4.8

1. Measure to the nearest centimeter.

____

____

Draw a line segment 6 centimeters long.

SRB 119–121

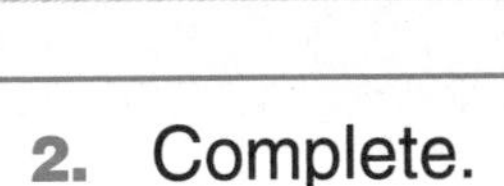

2. Complete.

_____ days in a week

_____ days in two weeks

_____ days in three weeks

_____ days in four weeks

3. Solve.

$2 \times 7 =$ _____

$8 \times 0 =$ _____

_____ $= 24 \times 1$

$5 \times$ _____ $= 50$

_____ $= 5 \times 5$

4. Complete.

20 dimes = $________

20 nickels = $________

20 quarters = $________

10 quarters and 7 dimes =

$________

5. Subtract. Show your work.

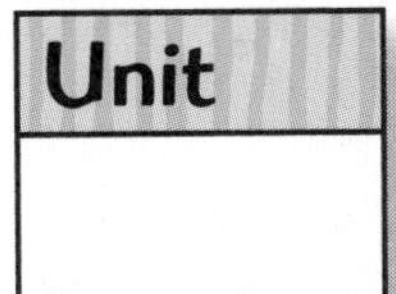

$$\begin{array}{r} 904 \\ -\ 368 \\ \hline \end{array} \qquad \begin{array}{r} 731 \\ -\ 53 \\ \hline \end{array}$$

SRB 54 55

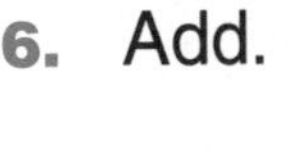

6. Add.

$15 + 15 + 13 =$ _____

$34 + 16 + 12 =$ _____

$23 + 13 + 17 =$ _____

$21 + 14 + 19 =$ _____

Date Time

Estimating Distances

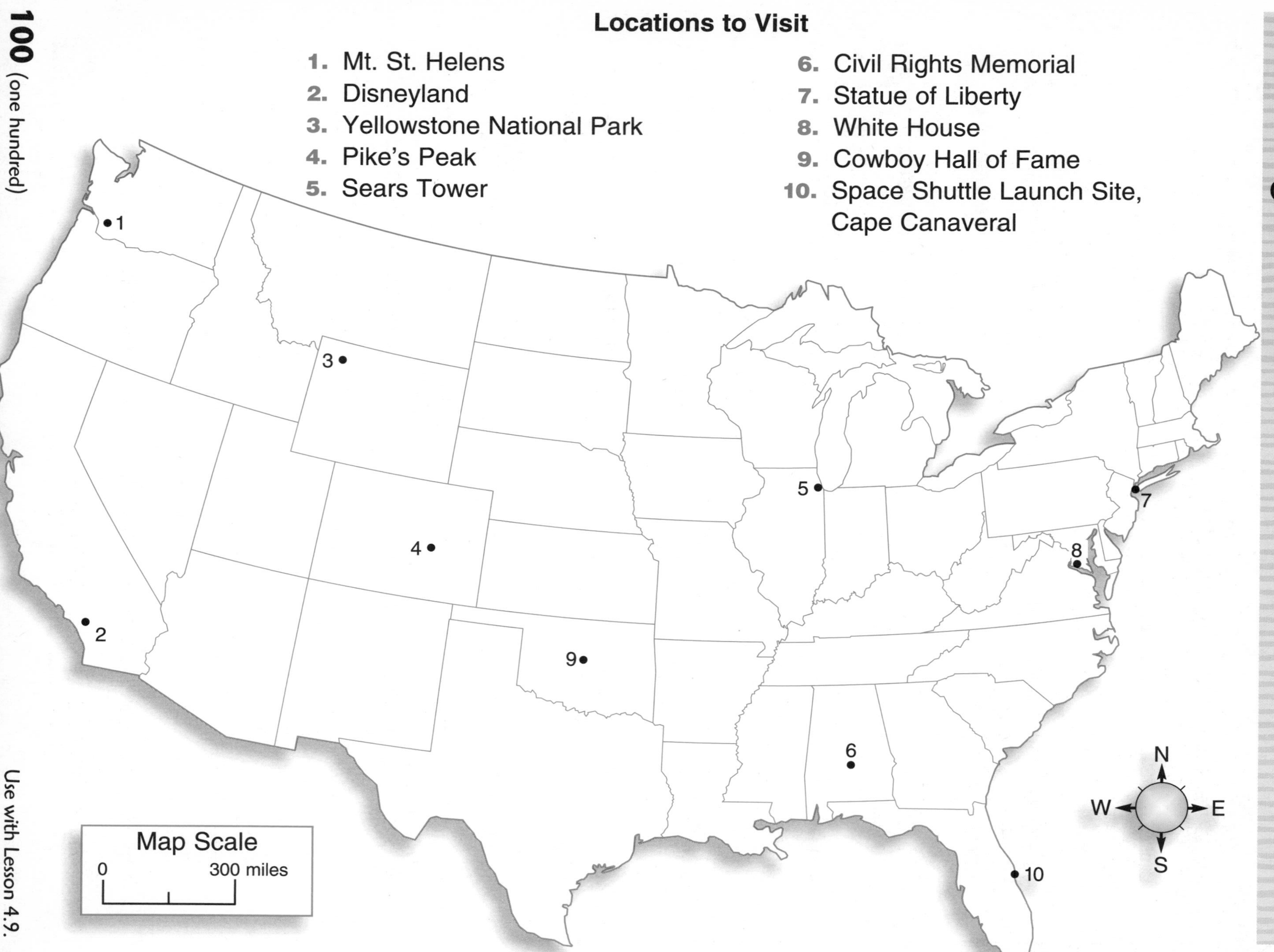

 Use with Lesson 4.9.

Date Time

A Pretend Trip

Pretend you want to take a trip to see some of the sights in the United States. Find out about how far it is between locations.

1. The Statue of Liberty is number __________.

   The Sears Tower is number __________.

   The distance between them is about __________ inches on the map.

   That is about __________ miles.

2. Pike's Peak is number __________.

   The White House is number __________.

   The distance between them is about __________ inches on the map.

   That is about __________ miles.

3. Yellowstone National Park is number __________.

   The Cowboy Hall of Fame is number __________.

   The distance between them is about __________ inches on the map.

   That is about __________ miles.

4. The Civil Rights Memorial is number __________.

   Disneyland is number __________.

   The distance between them is about __________ inches on the map.

   That is about __________ miles.

5. Make up one of your own.

   ____________________ is number __________.

   ____________________ is number __________.

   The distance between them is about __________ inches on the map.

   That is about __________ miles.

Date Time

Math Boxes 4.9

1. Use counters to solve.

18 marbles are shared equally.
Each child gets 5 marbles.
How many children are sharing?

_______________ (unit)

How many marbles are left over?

_______________ (unit)

2. Draw an array of 28 Xs arranged in 4 rows.

How many Xs in each row? _____

Write a number model for the array.

____

3. Solve.

$3 \times$ _____ $= 9$

_____ $= 4 \times 5$

$2 \times 6 =$ _____

$35 = 7 \times$ _____

$4 \times 6 =$ _____

$8 =$ _____ $\times 2$

4. Complete the Fact Triangle.
Write the fact family.

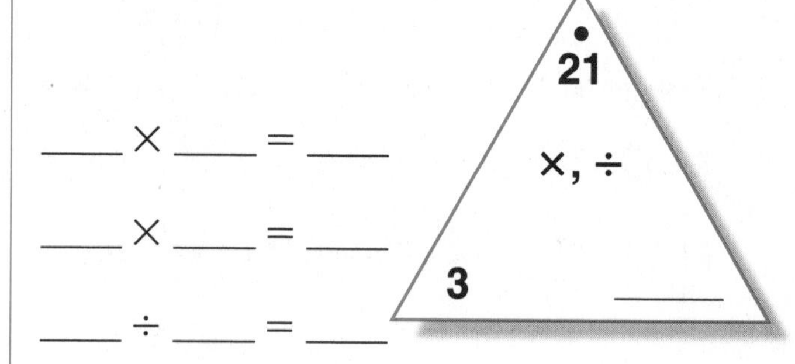

____ × ____ = ____

____ × ____ = ____

____ ÷ ____ = ____

____ ÷ ____ = ____

5. Solve. Each square equals 1 sq cm.

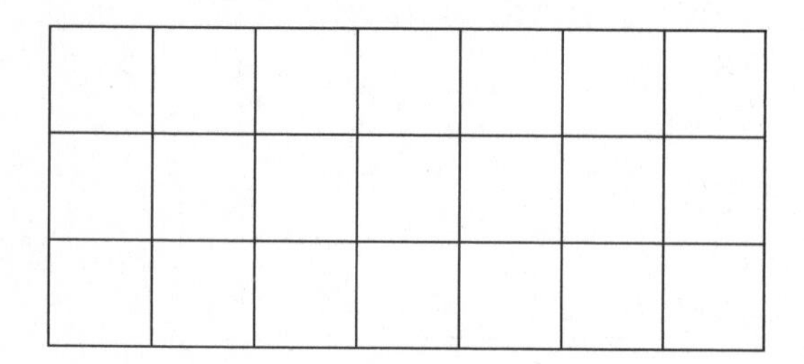

Area: _____ square centimeters

6. Complete.

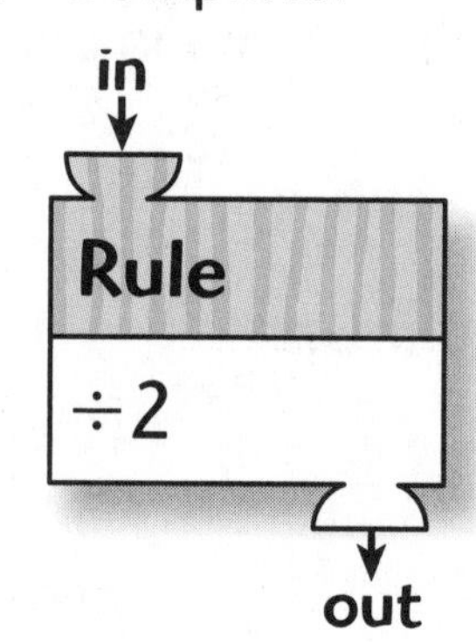

in	out
8	
16	
	10
50	

Date Time

Math Boxes 4.10

1. 56,937

Which digit is in the tens place? 3

Which digit is in the thousands place? ____

Which digit is in the hundreds place? ____

Which digit is in the ones place? ____

2. Put these numbers in order from smallest to largest.

4,073 47,003 43,700 7,430

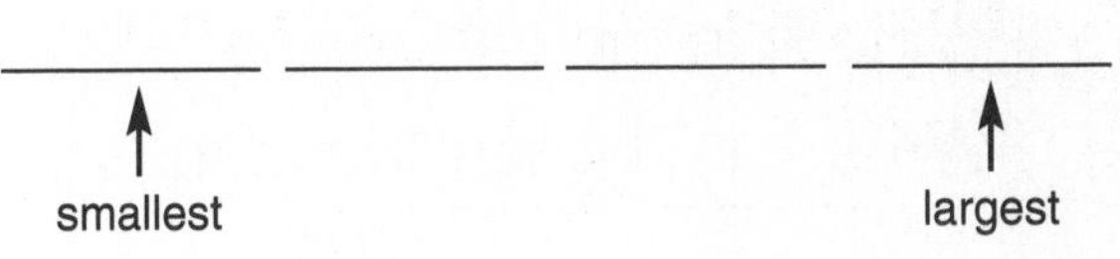

3. Write the number that has

5 hundreds
7 thousands
8 ones
4 tens
2 ten-thousands

____

Read it to a partner.

4. Which is more?

$3.45 or $3.09 ____________

$0.34 or $0.09 ____________

$14.50 or $14.55 ____________

5. Solve.

$$\begin{array}{r} 6{,}000 \\ 300 \\ 20 \\ +\quad 8 \\ \hline \end{array} \qquad \begin{array}{r} 400 \\ 9 \\ 8{,}000 \\ +\quad 30 \\ \hline \end{array}$$

6. Write the number that is 100 more.

76 ____________

300 ____________

471 ____________

8,634 ____________

5,925 ____________

Date Time

Place-Value Review

Follow the steps to find each number in Problems 1 and 2.

1. Write 6 in the ones place.
   Write 4 in the thousands place.
   Write 9 in the hundreds place.
   Write 0 in the tens place.
   Write 1 in the ten-thousands place.

   ____ ____ , ____ ____ ____

2. Write 6 in the tens place.
   Write 4 in the ten-thousands place.
   Write 9 in the ones place.
   Write 0 in the hundreds place.
   Write 1 in the thousands place.

   ____ ____ , ____ ____ ____

3. Compare the two numbers you wrote in Problems 1 and 2.

   Which is greater? ____________

4. Complete.

   The 9 in 4,965 stands for 9 hundreds or 900.

   The 7 in 87,629 stands for 7 ____________ or ____________.

   The 4 in 48,215 stands for 4 ____________ or ____________.

   The 0 in 72,601 stands for 0 ____________ or ____________.

Continue the counts.

5. 4,707; 4,708; 4,709; ________; ________; ________
6. 7,697; 7,698; 7,699; ________; ________; ________
7. 903; 902; 901; ________; ________; ________
8. 6,004; 6,003; 6,002; ________; ________; ________
9. 47,265; 47,266; 47,267; ________; ________; ________

Write the number that is 1,000 more.

10. 6,583 ________ 11. 9,990 ________ 12. 39,510 ________

Write the number that is 1,000 less.

13. 6,583 ________ 14. 9,990 ________ 15. 20,000 ________

Use with Lesson 5.1.

Date   Time

Math Boxes 5.1

1. 13 crayons are shared equally among 3 children.

How many crayons does each child get? ______________ (unit)

How many crayons are left over?

______________ (unit)

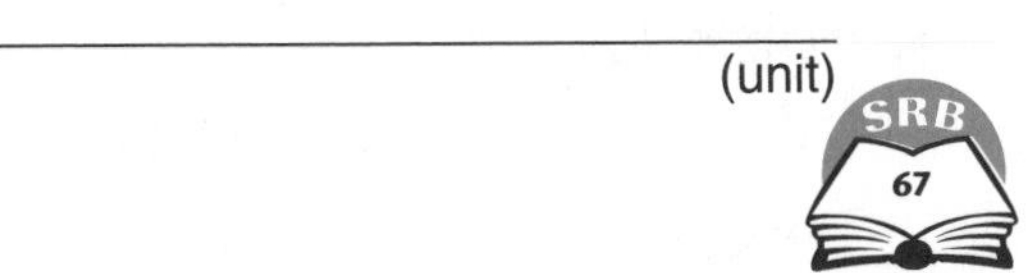

2. If a map scale shows that 1 inch represents 200 miles, then

2 inches represents ________ miles

3 inches represents ________ miles

5 inches represents ________ miles

7 inches represents ________ miles

3. Fill in the unit box. Then multiply.

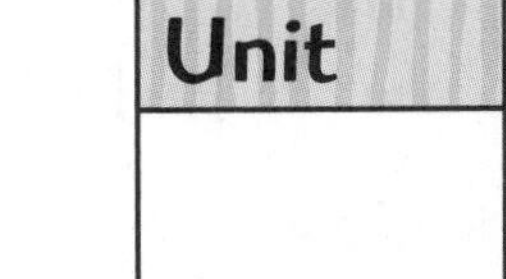

$2 \times 5 =$ _____

$7 \times 3 =$ _____

_____ $= 5 \times 5$

_____ $= 2 \times 7$

_____ $= 4 \times 6$

4. Complete the number-grid puzzles.

98

400

5. Draw a figure with a perimeter of 12 centimeters.

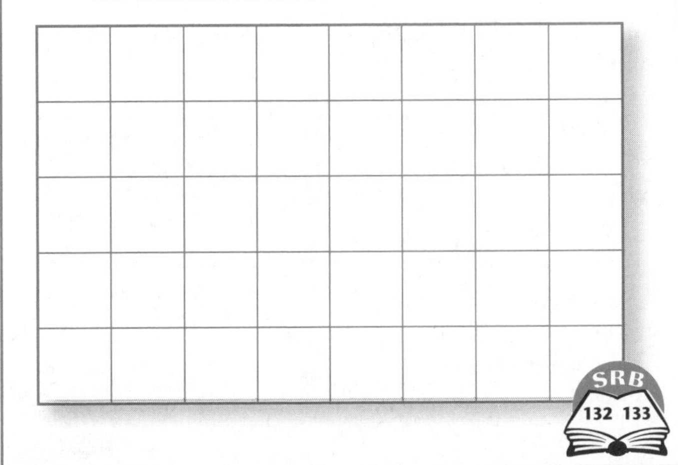

SRB 132 133

6. The "about 3 times" circle rule:
For any circle, the circumference is about 3 times the diameter.

Unit
inches

diameter	circumference
8	
10	
50	

SRB 134 135

Date Time

Math Boxes 5.2

1. Write the number. This number has

   7 thousands

   8 tens

   5 ten-thousands

   1 one

   0 hundreds

   ___ ___, ___ ___ ___

2. Complete the Fact Triangle and write the fact family.

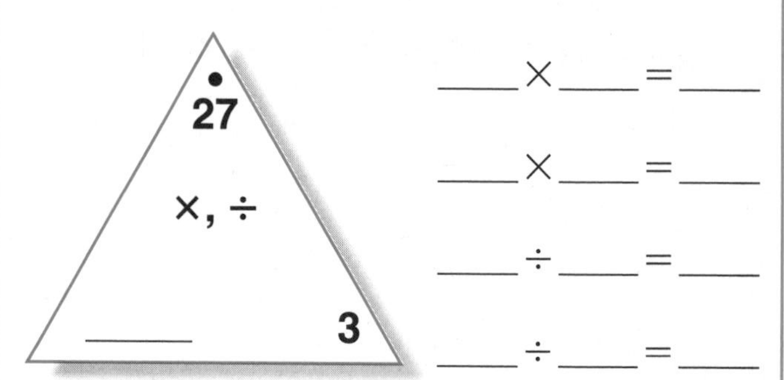

___ × ___ = ___

___ × ___ = ___

___ ÷ ___ = ___

___ ÷ ___ = ___

3. Draw a 4 × 6 rectangle.

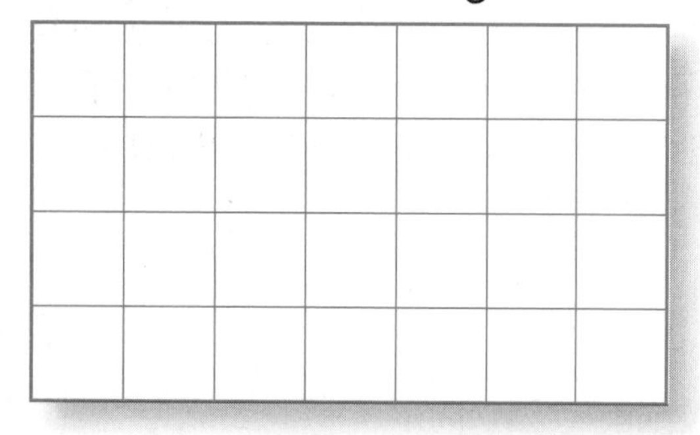

Number model: ___ × ___ = ___

Area: _____ square units

4. Write a multiplication story by filling in the blanks.

   8 rows.

   5 ______________ in each row.

   How many in all? ______

   Write a number model.

   ____

5. Fill in the rule and then the empty frames.

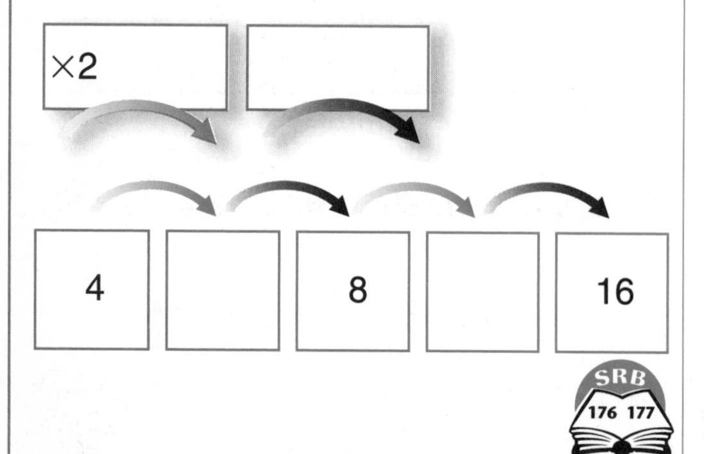

6. Fill in the unit box. Write the missing number in the diagram. Write a number model.

   ____

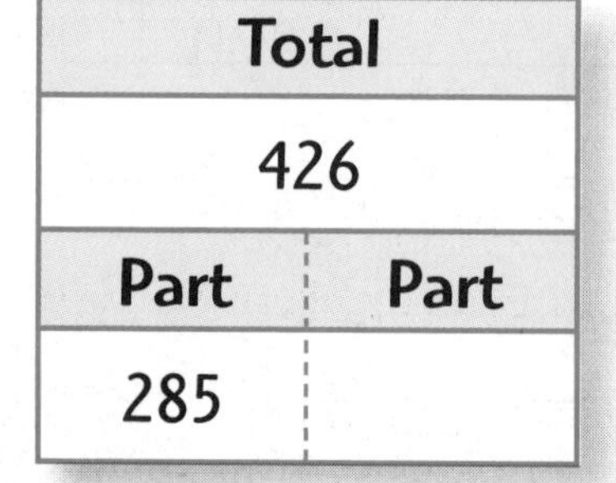

Total	
426	
Part	Part
285	

Unit

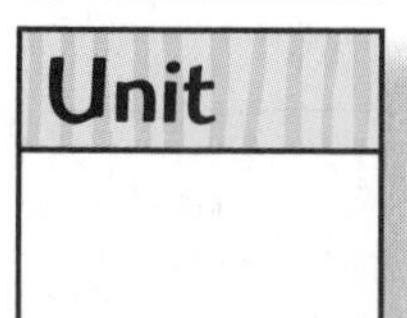

Math Boxes 5.3

1. Write <, >, or =.

263,473 _____ 263,107

37,261 _____ 37,621

99,999 _____ 111,111

Make up your own.

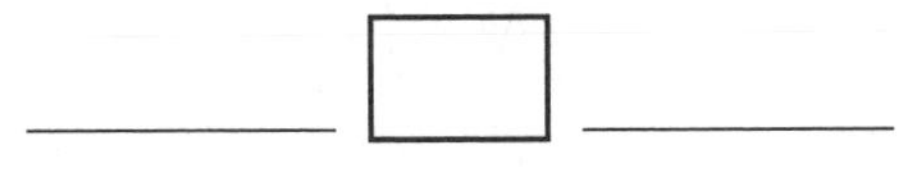

2. If a map scale shows that 1 cm represents 1,000 km, then

2 cm represents __________ km

9 cm represents __________ km

16 cm represents __________ km

20 cm represents __________ km

3. Fill in the unit box. Then multiply.

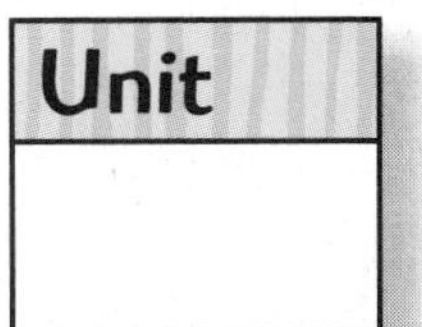

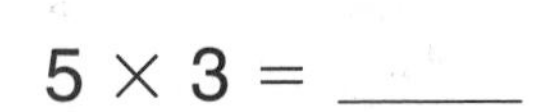

$5 \times 3 =$ _____

_____ $= 4 \times 5$

$3 \times 3 =$ _____

_____ $= 7 \times 3$

_____ $= 5 \times 5$

SRB 46 47

4. On Tuesday Gabriela put $76 in her bank account. Now she has $123. How much money did she have in her bank account on Monday?

$__________

5.

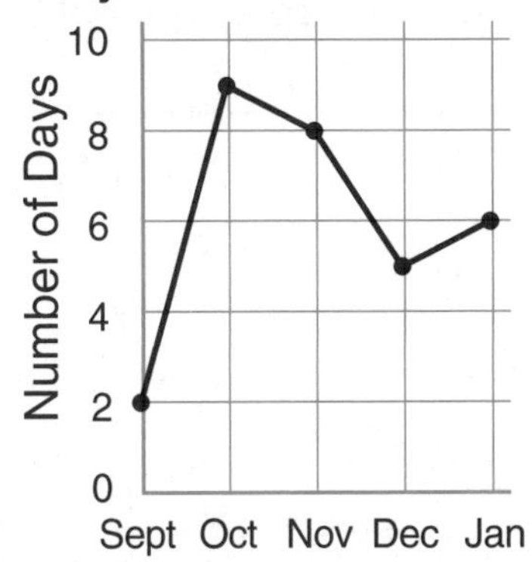

What is the median number of days of indoor recess? ____ days

SRB 74 82

6. Measure to the nearest $\frac{1}{4}$ inch.

____

Draw a line segment $2\frac{1}{4}$ inches long.

Date Time

Working with Populations

10 U.S. Cities with the Largest Populations

	1980*	1995*
New York, NY	7,071,639	7,380,906
Los Angeles, CA	2,968,528	3,553,638
Chicago, IL	3,005,072	2,721,547
Houston, TX	1,611,382	1,744,058
Philadelphia, PA	1,688,210	1,478,002
San Diego, CA	875,538	1,171,121
Phoenix, AZ	790,183	1,159,014
San Antonio, TX	785,940	1,067,816
Dallas, TX	1,007,618	1,053,292
Detroit, MI	1,027,974	1,000,272

*U.S. Census data

Use this table to solve the problems.

1. List the cities that gained population from 1980 to 1995.

____

____

2. List the cities that lost population from 1980 to 1995.

____

3. Look at your answers to Problem 1. Name a city where the population increased by

   a. more than 100,000 b. about 100,000 c. less than 100,000

   _______ _______ _______

4. In 1980, which two cities had a population about half that of Houston, TX? _______________

5. In 1995, which city had a population about double that of Philadelphia, PA? _______________

6. Which city had the smallest change in population? _______________

Date Time

Math Boxes 5.4

1. For the number 5,749,862

the 4 means 40,000

the 5 means ______

the 8 means ______

the 7 means ______

the 9 means ______

2. Complete the Fact Triangle and write the fact family.

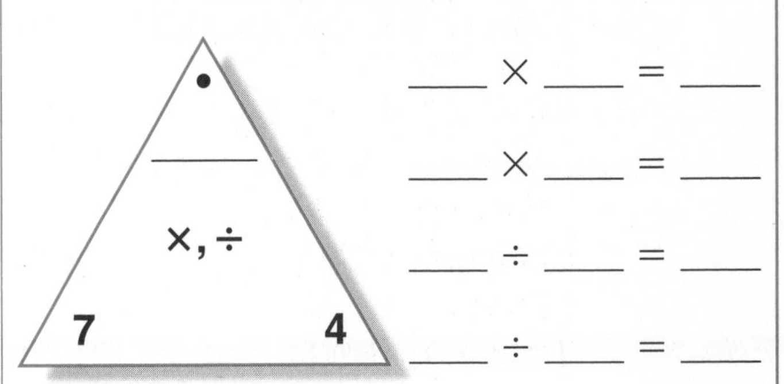

___ × ___ = ___

___ × ___ = ___

___ ÷ ___ = ___

___ ÷ ___ = ___

3. Find the perimeter.

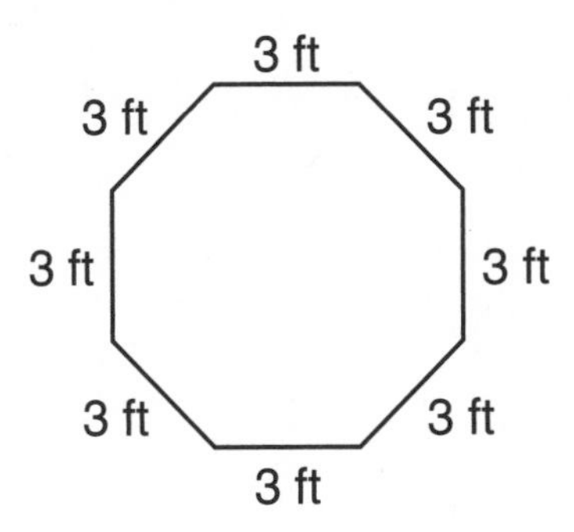

Perimeter = ______ (unit)

4. Teesha has 345 marbles. Keiko has 279 marbles. How many fewer marbles does Keiko have than Teesha?

______ marbles

5. Write a division story by filling in the blanks.

There are 48 ______ in 6 rows.

How many ______ are in each row? ____

Write a number model.

______

6. Measure to the nearest centimeter.

______

______

Draw a line segment 4.5 centimeters long.

SRB 119–121

Date Time

How Old Am I?

1. On what date were you born? __________

2. How old were you on your last birthday? __________ years old

3. About how many minutes old do you think you were on your last birthday? Make an X next to your guess.

_____ between 10,000 and 100,000 minutes

_____ between 100,000 and 1,000,000 minutes

_____ between 1,000,000 and 10,000,000 minutes

Use your calculator.

4. a. About how many days old were you on your last birthday? Do not include any leap year days. ____________________

   b. That's about how many hours? ____________________

   c. That's about how many minutes? ____________________

Challenge

Adding Leap Year Days

5. a. List all of the leap years from the time you were born to your last birthday. ____________________

   b. That adds how many extra days to your last birthday? ____________________

   c. How many extra minutes? ____________________

6. Add the number of extra minutes to the number of minutes in your answer in Problem 4c. How many minutes are there in all? ____________________

7. On my last birthday, I was about ____________________ minutes old.

Date Time

Math Boxes 5.5

1. Circle the largest number.
Underline the smallest number.

1,099,999

697,432

697,500

697,490

1,110,000

697,433

2. If a map scale shows that 1 in. represents 50 miles, then

_____ in. represents 200 miles

_____ in. represents 300 miles

9 in. represents _________ miles

11 in. represents _________ miles

3. Circle the number that is about 10,000 less than 30,000.

56,023 21,004

35,900 15,999

SRB 18 19

4. Fill in the unit box. Then multiply.

$4 \times 3 =$ _____

$2 \times 7 =$ _____

_____ $= 5 \times 7$

_____ $= 2 \times 5$

_____ $= 6 \times 5$

5. Body-plus-tail lengths (inches) for 13 cats:

30, 29, 28, 24, 29, 35, 16, 27, 29, 36, 28, 31, 32

Median = _____

Maximum = _____

6. Draw a shape with an area of 16 square units.

How many sides does your shape have? _____ sides

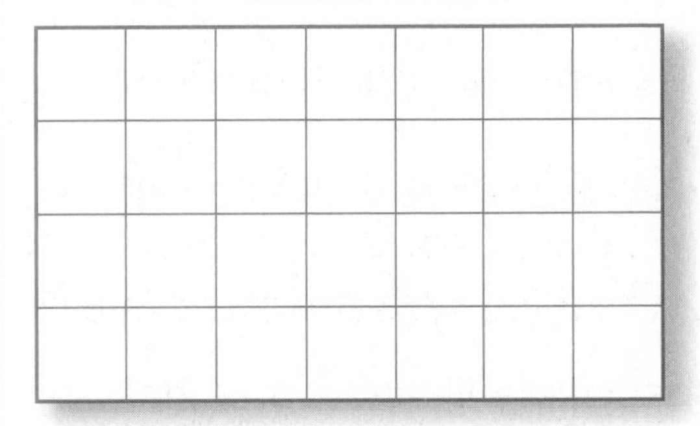

Date Time

Finding the Value of Base-10 Blocks

Materials ☐ classroom supply of base-10 blocks

Work in a group.

1. Estimate the value of the base-10 blocks. Don't let anyone in your group see your estimate.

   Estimate: ________________

2. Plan how your group will find the actual value of the blocks and what each person will do to help. Then carry out your plan. Describe your job.

   ____

   ____

3. What is the actual value of the base-10 blocks? ________________

4. Write the estimates of your group and the actual value of the base-10 blocks in order from smallest to largest. Circle the actual value of the base-10 blocks.

   ____

   ____

5. **a.** Which estimate was closest to the actual value? __________

   b. How many estimates were higher than the closest estimate? __________

   c. How many estimates were lower than the closest estimate? __________

   d. How far was the highest estimate from the actual value? __________

   e. How far was the lowest estimate from the actual value? __________

6. How does your estimate compare to the actual value? __________

7. If you have extra time, put part of the block supply to the side.

   First estimate its value and then find its actual value.

Date Time

Squares, Rectangles, and Triangles

Materials ❑ straightedge

A

H • • E

D • • B

G • • F

C

Work on your own or with a partner.

1. Use your straightedge to draw line segments between points *A* and *B*, *B* and *C*, *C* and *D*, and *D* and *A*.

   What kind of shape did you draw? ______________________

2. Now draw line segments between points *E* and *F*, *F* and *G*, *G* and *H*, and *H* and *E*.

   What kind of shape did you draw? ______________________

3. Draw line segments between points *E* and *G* and between points *F* and *H*.

   How many different sizes of squares are there? ______________________

   How many squares in all? ______________________

4. How many different sizes of triangles are there? ______________________

   How many triangles in all? ______________________

5. How many rectangles are there that are not squares? ______________________

Date Time

Pattern-Block Perimeters

Materials ❑ pattern blocks: square, large rhombus, small rhombus, triangle

Work on your own or with a partner.

1. Imagine that each polygon is "rolled" along a line, starting at point *S*. Estimate the distance each polygon will "roll" after 1 full turn. Mark an X at the point you think the polygon will reach.

2. Check your estimate by "rolling" a pattern block that matches the polygon. Circle the point reached by the pattern block.

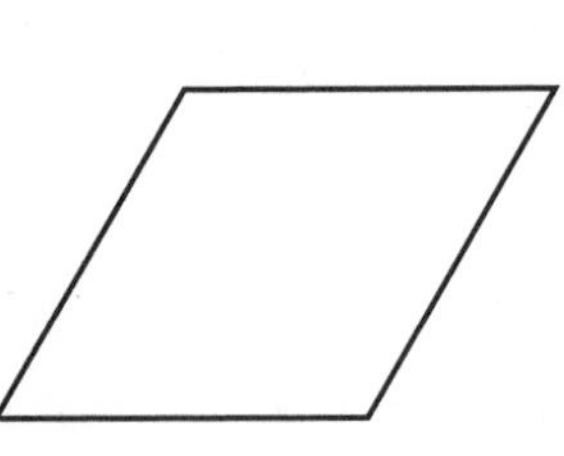

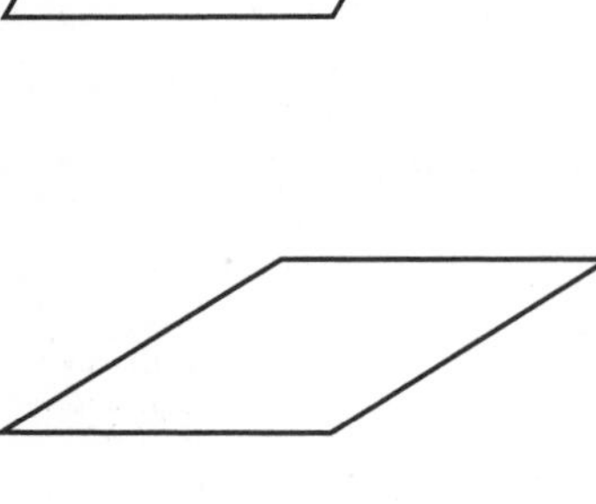

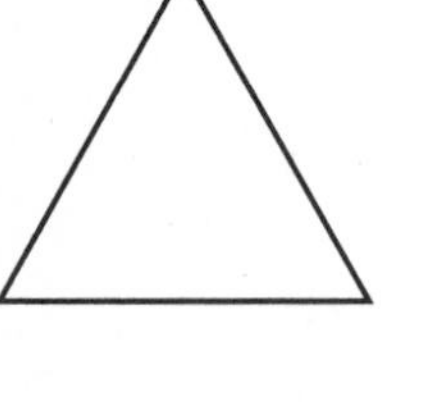
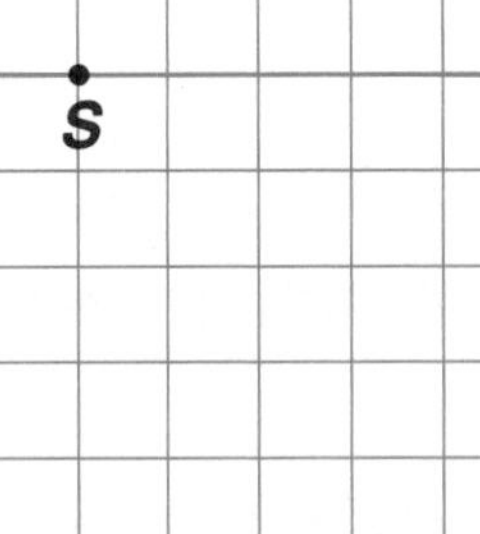
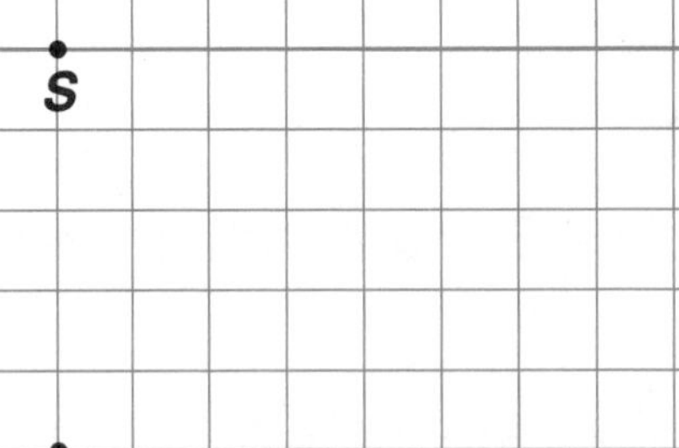
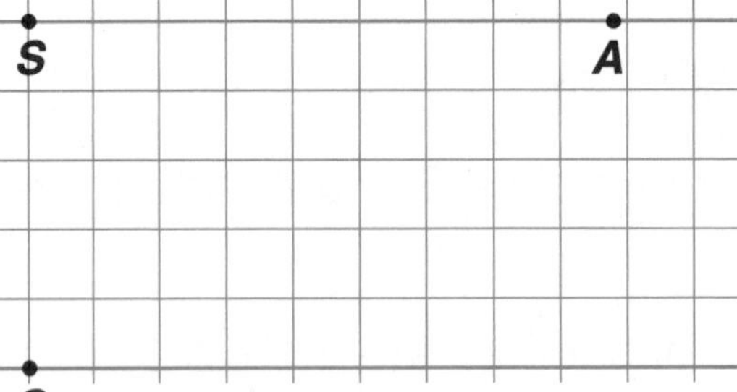

3. Which 3 shapes have about the same perimeter?

____

4. Which of these 3 shapes do you think has the largest area? ____________

5. Which of the 4 shapes do you think has the smallest area? ____________

Date Time

Math Boxes 5.6

1. In the number 6,940,173

the 9 means 900,000

the 6 means ______

the 1 means ______

the 4 means ______

the 7 means ______

2. Complete the number-grid puzzle.

8,742

SRB 7 8

3. Use your calculator.

Enter	Change to	How?
894	12,894	
1,366	966	
627,581	628,581	
43,775	43,175	

SRB 18 19

4. Draw a 7-by-6 array.

What is the number model?

5. Fill in the rule and then fill in the empty frames.

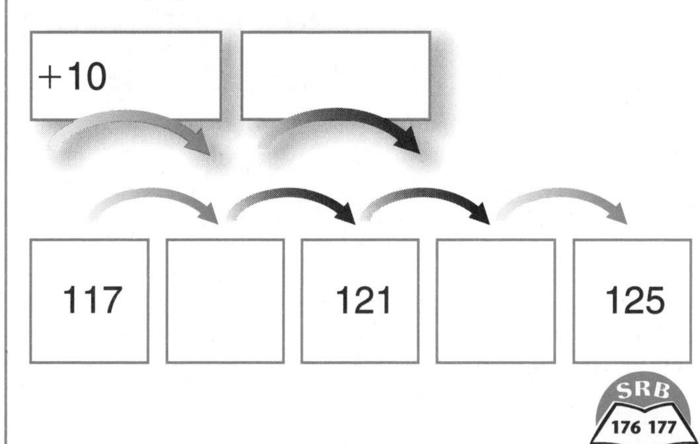

SRB 176 177

6. Add.

$$\begin{array}{r} 28 \\ 374 \\ +\ 101 \\ \hline \end{array} \qquad \begin{array}{r} 72 \\ 407 \\ 283 \\ +\ 19 \\ \hline \end{array}$$

SRB 51–53

Date Time

Place Value in Decimals

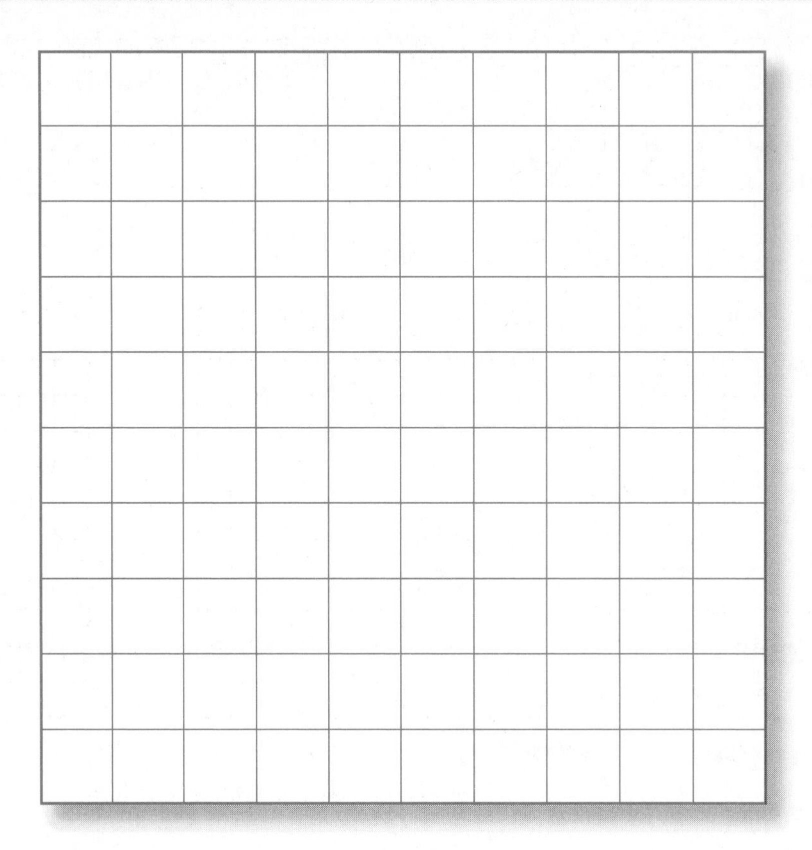

If the grid is ONE, then which part of each grid is shaded?

Write a decimal and a fraction below each grid.

1.

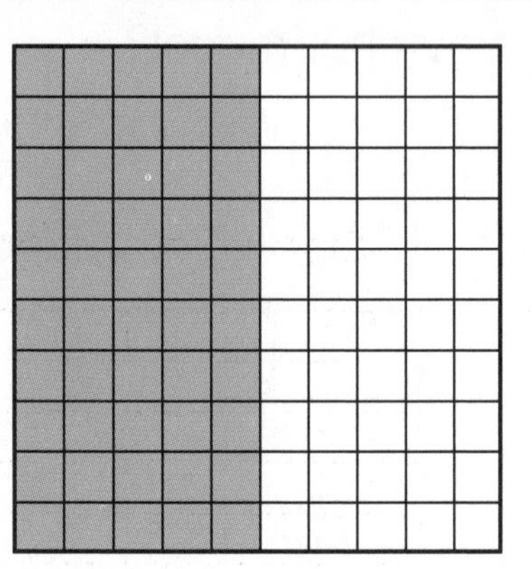

fraction: ____________

decimal: ____________

2.

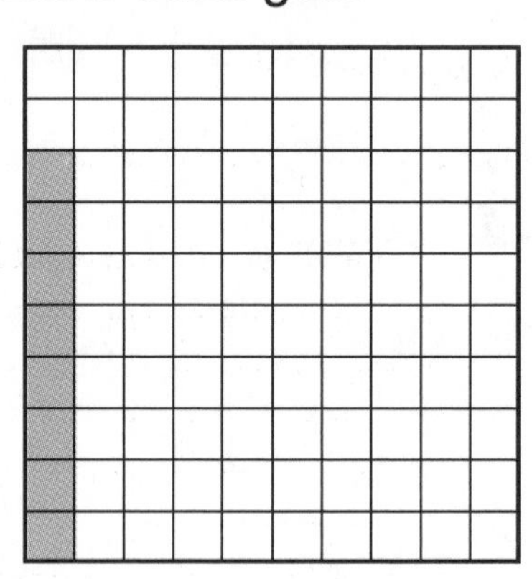

fraction: ____________

decimal: ____________

3.

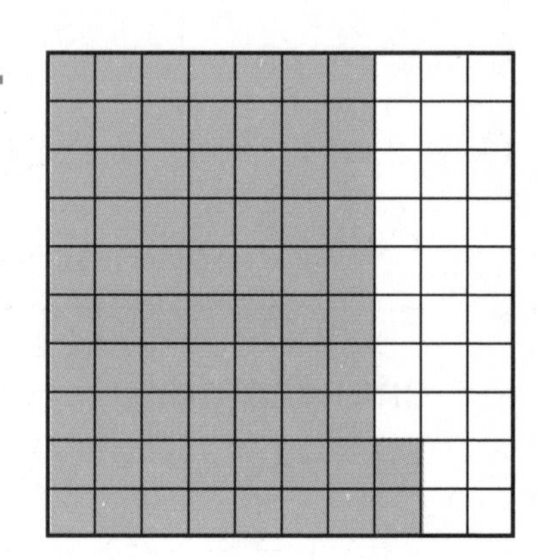

fraction: ____________

decimal: ____________

Place Value in Decimals (cont.)

4. Which decimal in each pair is greater? Use the grids in Exercises 1–3 to help you.

0.5 or 0.08 __________ 0.08 or 0.72 __________ 0.5 or 0.72 __________

Color part of each grid to show the decimal named.

5. Color 0.7 of the grid.

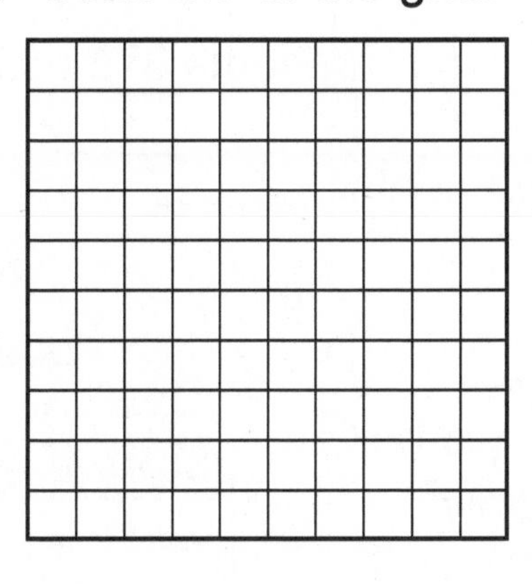

6. Color 0.07 of the grid.

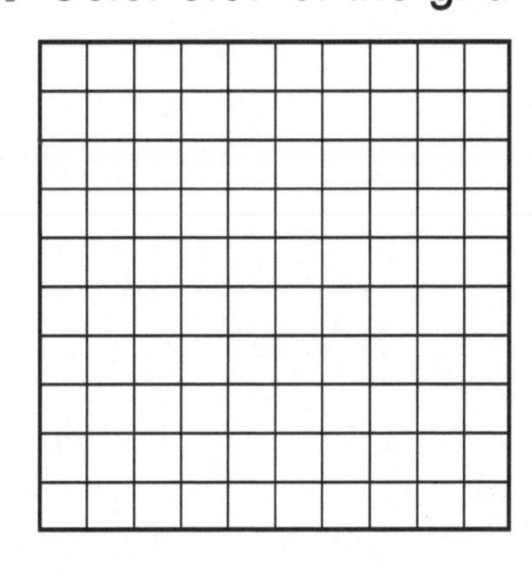

7. Color 0.46 of the grid.

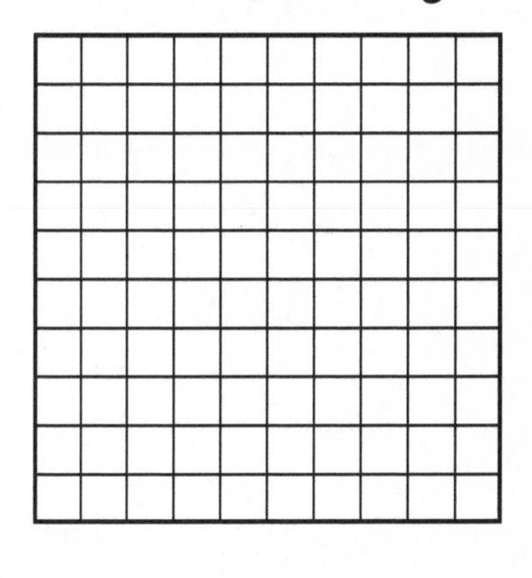

8. Write 0.7, 0.07, and 0.46 in order from smallest to largest.

Use the grids in Exercises 5–7 to help you. __________ __________ __________

Challenge

Color part of each grid to show the fraction named.

9. Color $\frac{4}{10}$ of the grid.

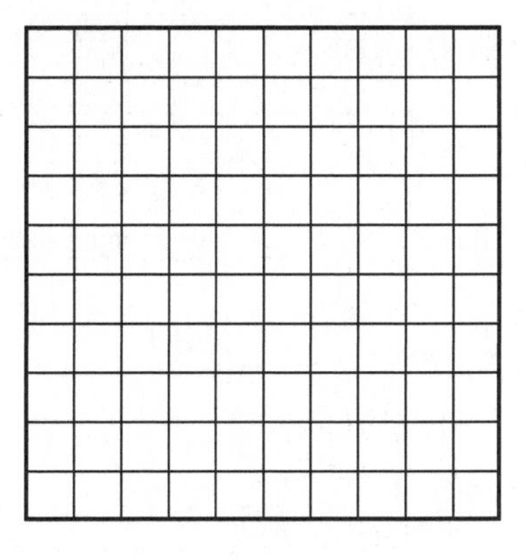

10. Color $\frac{1}{2}$ of the grid.

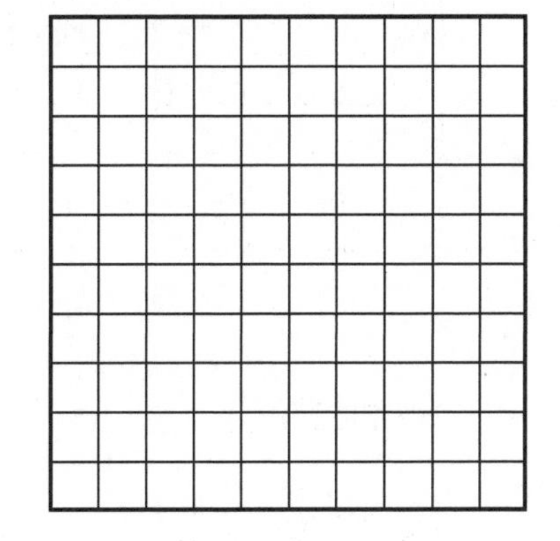

11. Color $\frac{23}{100}$ of the grid.

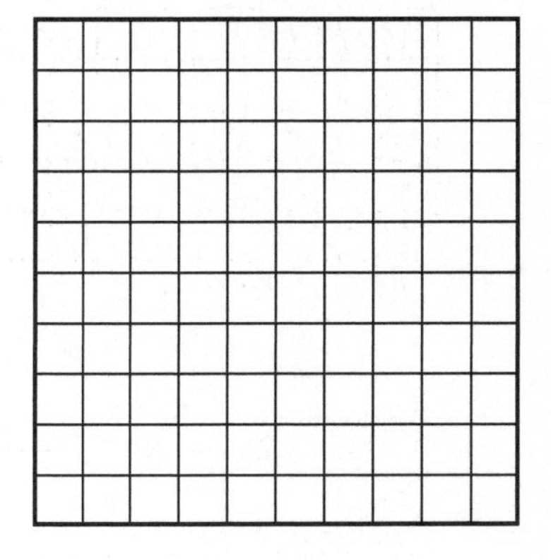

12. Write $\frac{23}{100}$ as a decimal. __________

Date Time

Math Boxes 5.7

1. Circle the largest number. Underline the smallest number.

2,999,999

946,487

946,800

946,793

4,000,007

946,200

2. If a map scale shows that 1 cm represents 25 miles, then

_______ cm represents 125 miles

_______ cm represents 200 miles

_______ cm represents 375 miles

20 cm represents _______ miles

3. Solve.

Double 2 ___________

Double 10 ___________

Double 75 ___________

Double 1,000 _________

Double 1,500 _________

4. Fill in the unit box. Then multiply.

Unit

$4 \times 5 =$ _____

$2 \times 6 =$ _____

$3 \times 5 =$ _____

_____ $= 7 \times 4$

_____ $= 6 \times 5$

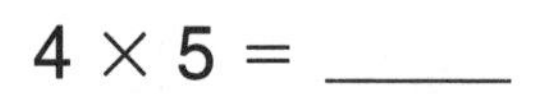

5. Ages of 9 teachers:

30, 24, 49, 50, 38, 44, 40, 35, 51

Median = _____

Maximum = _____

6. Find the perimeter.

Unit
yards

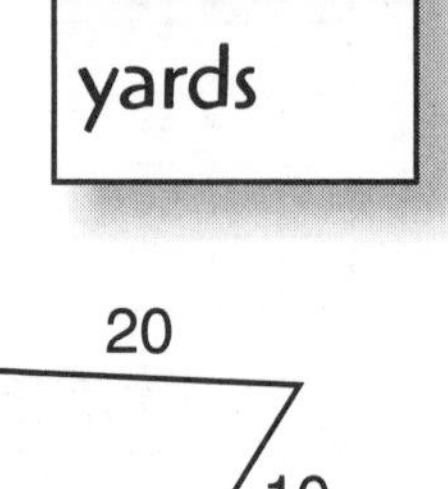

Perimeter: ___________ (unit)

Date Time

Exploring Decimals

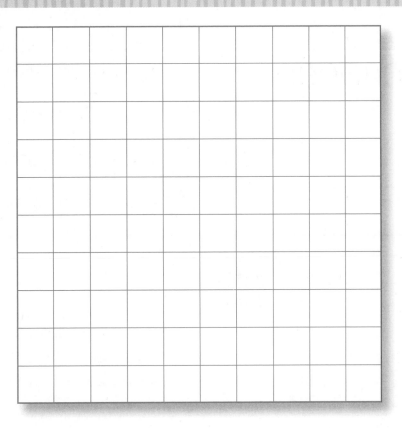

A	B	C	D
27 hundredths	2 tenths, 7 hundredths	0. 27	$\frac{27}{100}$
______ hundredths	______ tenths, ______ hundredths	0. ______	
______ hundredths	______ tenths, ______ hundredths	0. ______	
______ hundredths	______ tenths, ______ hundredths	0. ______	
______ hundredths	______ tenths, ______ hundredths	0. ______	
______ hundredths	______ tenths, ______ hundredths	0. ______	
______ hundredths	______ tenths, ______ hundredths	0. ______	

Date Time

Math Boxes 5.8

1. For the number 4,963,521

4 means 4,000,000

3 means ______

1 means ______

6 means ______

9 means ______

2. If each grid is ONE, what part of each grid is shaded? Write the decimal.

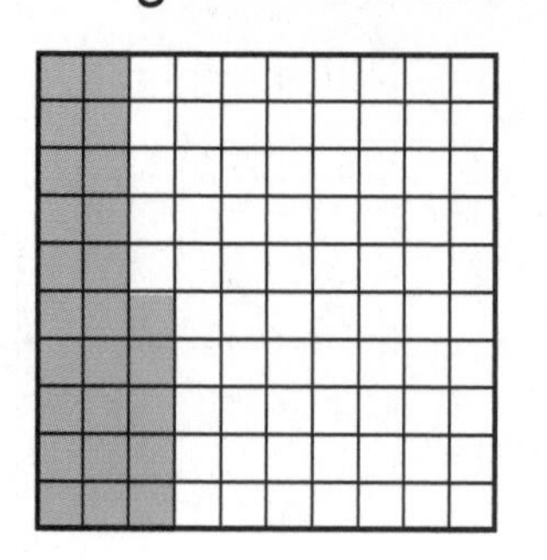

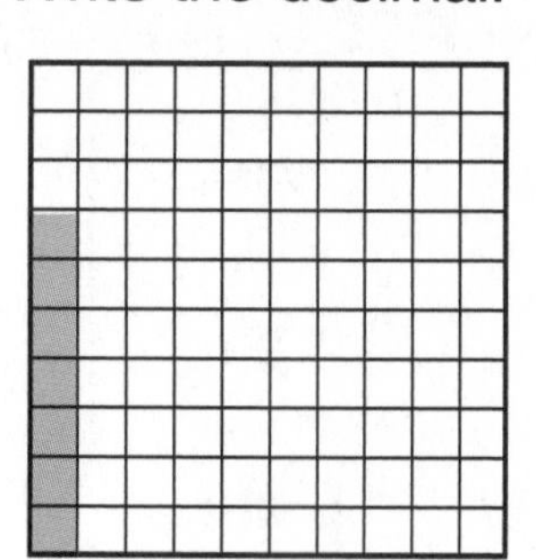

______ ______

SRB 33 34

3. Solve.

$$\begin{array}{r} 3,976 \\ -\ 1,439 \\ \hline \end{array}$$

$$\begin{array}{r} 6,840 \\ -\ 5,695 \\ \hline \end{array}$$

$$\begin{array}{r} 14,256 \\ -\ 3,661 \\ \hline \end{array}$$

4. How many slices does each person get if 64 slices of pizza are shared equally among 4 people?

Answer: ______ (unit)

Number model:

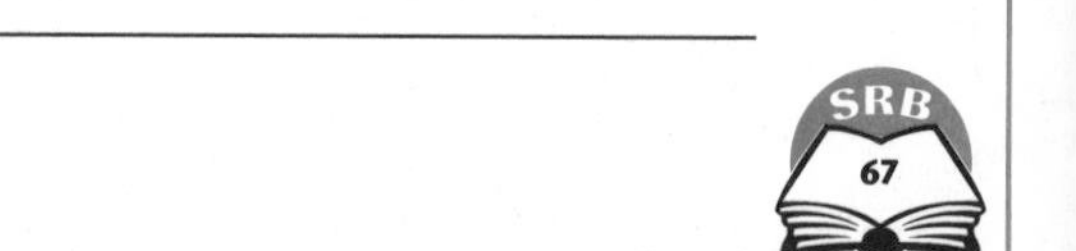

5. Draw a 3 × 7 rectangle.

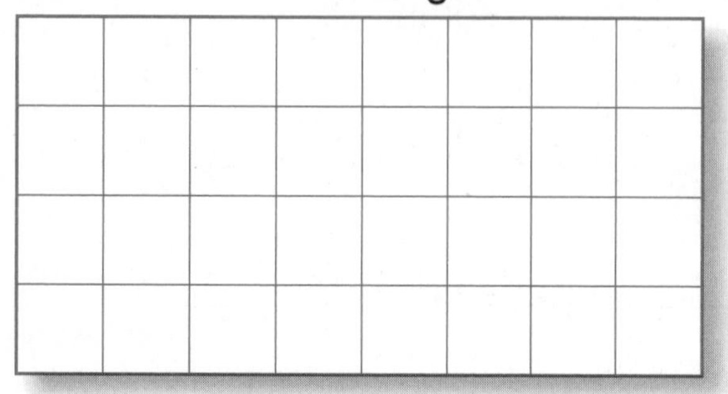

Number model: ___ × ___ = _____

Area: _____ square units

6. Draw the hands to show 5:50.

How long until 8:30?

_____ hours _____ minutes

Date Time

Decimals for Metric Measurements

1. Fill in the missing information. Put longs and cubes end to end on a meterstick to help you.

Length in Centimeters	Number of Longs	Number of Cubes	Length in Meters
24 cm	2	4	0.24 m
36 cm	______	______	______ m
______ cm	0	3	______ m
8 cm	______	______	______ m
______ cm	______	______	0.3 m
______ cm	4	3	______ m

Work with a partner. Each partner uses base-10 blocks to make one length in each pair. Compare the lengths and circle the one that is greater.

2. 0.18 or 0.5
3. 0.2 or 0.08
4. 0.09 or 0.12
5. 0.24 or 0.42
6. 0.10 or 0.02
7. 0.3 or 0.24

Follow these directions on the ruler below.

8. Make a dot at 4 cm and label it with the letter *A*.
9. Make a dot at 0.1 m and label it with the letter *B*.
10. Make a dot at 0.15 m and label it with the letter *C*.
11. Make a dot at 0.08 m and label it with the letter *D*.

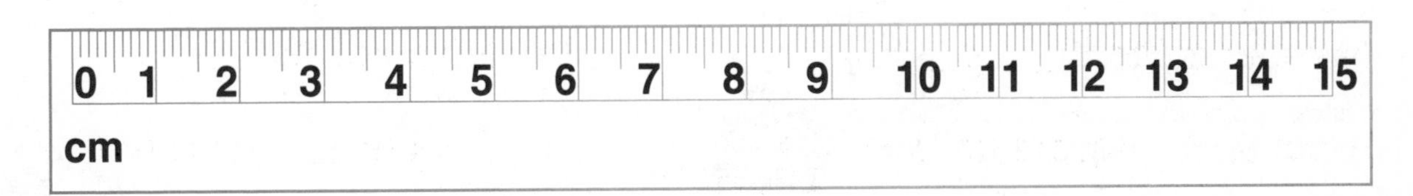

Date Time

Math Boxes 5.9

1. Put these numbers in order from smallest to largest.

998,752 __________

1,000,008 __________

750,999 __________

1,709,832 __________

2. Write the number that has

2 in the ones place

6 in the tenths place

7 in the hundredths place

____ . ____ ____

3. Solve.

Double 6 __________

Double 24 __________

Double 59 __________

Double 113 __________

Double 642 __________

4. Fill in the unit box. Then multiply.

Unit

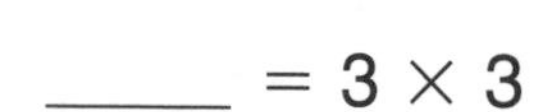

____ $= 3 \times 3$

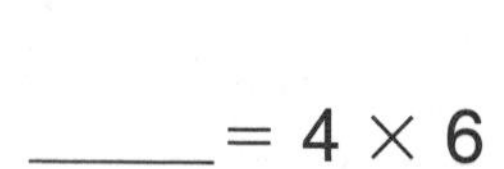

____ $= 4 \times 6$

$5 \times 5 =$ ____

$3 \times 6 =$ ____

$2 \times 4 =$ ____

5. Median number of books read? ____

Maximum number of books read? ____

Books Read

0 1 2 3 4 5 6

Jen Mark Inez Lisa Joe

6. 7 boxes. 7 cans per box.

How many cans in all?

________ cans

9 cars. 3 people per car.

How many people in all?

________ people

Date Time

How Wet? How Dry?

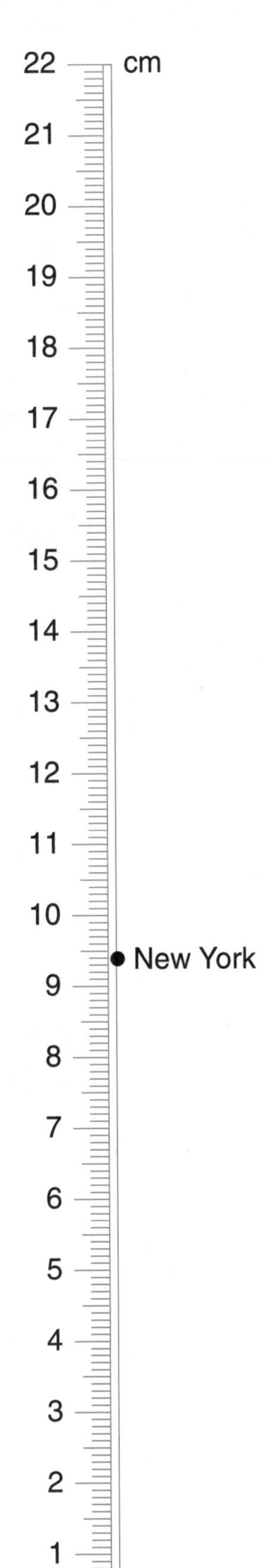

1. Use the scale at the left and the map on page 245 of the *Student Reference Book.* Make a dot for the level of precipitation in each of the following cities: Phoenix, Helena, Denver, Cleveland, and Asheville. Write the name of the city next to the dot.

2. Which city gets about 2 centimeters less rain than New York?

   ____

3. Which city gets about half as much rain as Denver?

   ____

4. Which city gets about 5 times as much rain as Helena?

   ____

5. A decimeter is 10 centimeters. Which cities on the map get at least 1 decimeter of rain?

   ____

   ____

Did You Know?

According to the National Geographic Society, the rainiest place in the world is Mount Waialeale in Hawaii. It rains an average of about 1,170 centimeters a year on Mount Waialeale.

Challenge

6. Suppose it rained 1,170 centimeters in your classroom. Would the water reach the ceiling?

   __________ millimeters = 1,170 centimeters = __________ meters

   Answer: ______________

Math Boxes 5.10

1. Complete the Frames and Arrows.

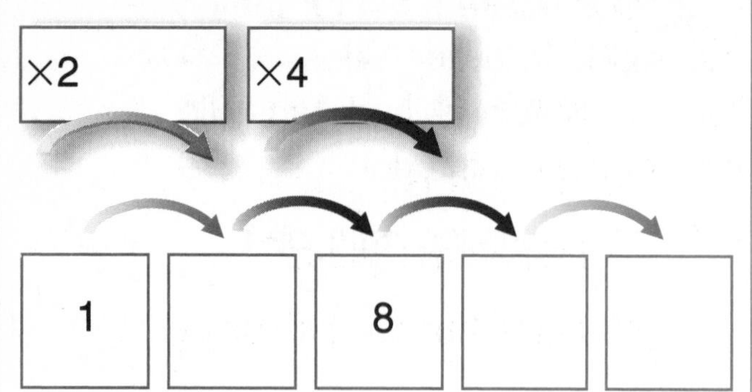

SRB 176 177

2. Color 0.6 of the grid.

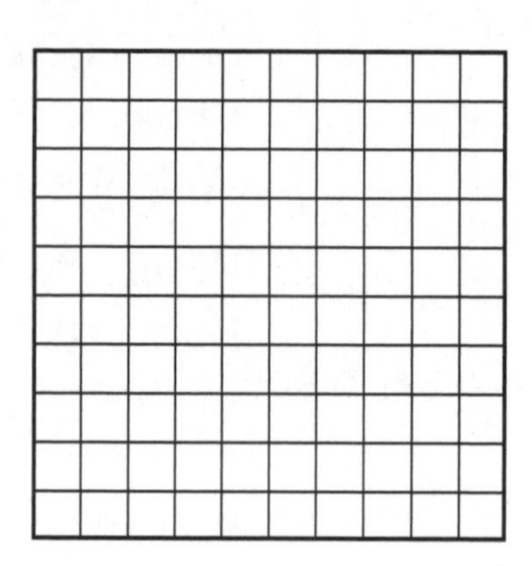

SRB 33 34

3. Complete.

2 hours = _______ minutes

5 weeks = _______ days

3 hours = _______ minutes

2 years = _______ days

4. Circle any measurements in Column B that match the one in Column A.

Column A	Column B	
2 feet	12 in.	3 yd
	24 in.	1 yd
3 feet	36 in.	1 m
	1 yd	30 in.
2 yards	50 in.	72 in.
	6 ft	9 ft

5. Add.

$$\begin{array}{r} 3 \\ 96 \\ 104 \\ +\ 327 \\ \hline \end{array} \qquad \begin{array}{r} 3 \\ 33 \\ 333 \\ +\ 3{,}333 \\ \hline \end{array}$$

6. Complete.

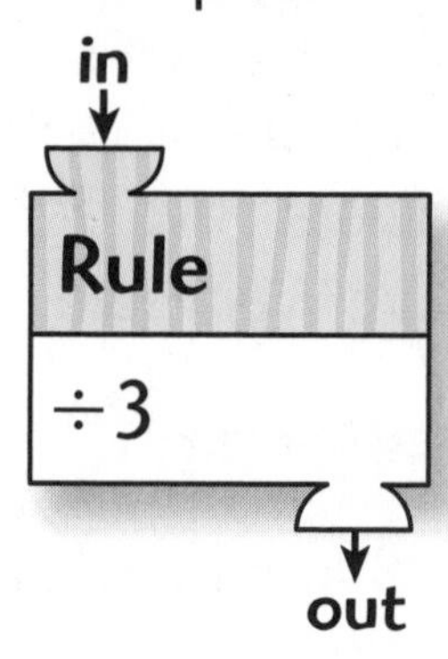

in	out
9	
15	
	7
	10

Date Time

More Decimals

Use your place-value tool to help you.

Write the number that matches each description.

1. 4 in the tenths place
   2 in the thousandths place
   7 in the hundredths place
   0 in the ones place
   ____

2. 5 in the tenths place
   3 in the tens place
   5 in the ones place
   3 in the hundredths place
   ____

3. 4 in the thousandths place
   2 in the ones place
   7 in the hundredths place
   0 in the tenths place
   ____

4. 0 in the hundredths place
   6 in the ones place
   8 in the thousandths place
   0 in the tenths place
   ____

5. With your partner, decide how to read each of the decimals in Problems 1–4.

Write each number below as a decimal.

6. nine-tenths ________
7. thirty-thousandths ________
8. fifty-three hundredths ________
9. sixty and four-tenths ________
10. seven and seven-thousandths ________
11. sixty and four-hundredths ________
12. eight hundred ________
13. sixty-two thousandths ________

Fill in the missing numbers.

Unit
meter

14.

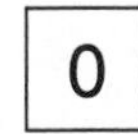

0 ____ ____ ____ ____ ____ ____ ____ ____ ____ 1

15.

0 ____ ____ ____ ____ ____ ____ ____ ____ ____ 0.1

Date Time

Math Boxes 5.11

1. How much of this grid is shaded?

____ • ____ ____

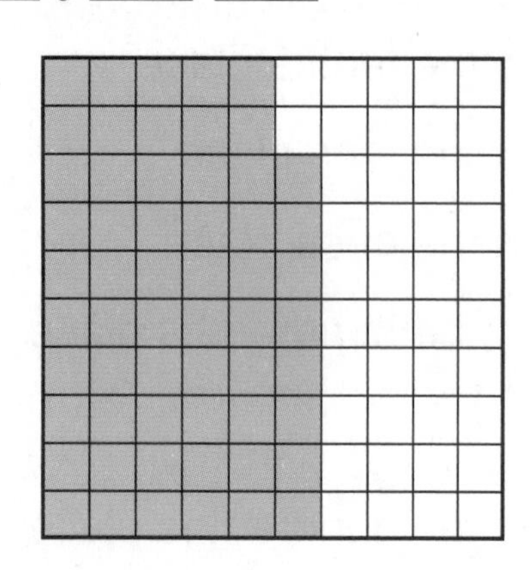

SRB 33 34

2. Write the number that has

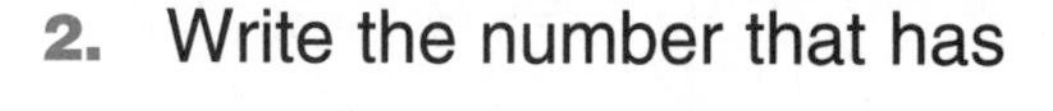

4 in the tenths place

0 in the hundredths place

6 in the ones place

9 in the thousandths place

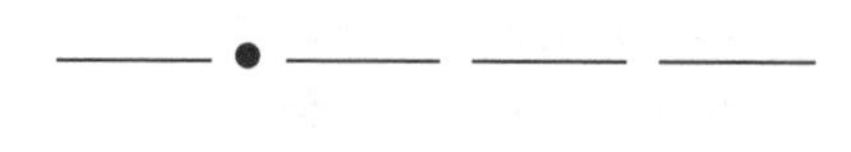

____ • ____ ____ ____

3. Circle the number that is about 1 million less than 6 million.

50,023

6,900,800

4,986,500

3,090,222

4. Fill in the unit box. Then multiply.

Unit

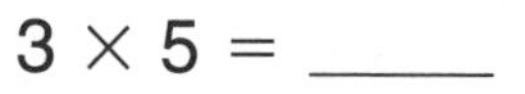

$3 \times 5 =$ ____

$4 \times 6 =$ ____

____ $= 7 \times 5$

____ $= 4 \times 4$

____ $= 6 \times 3$

5. Draw a 4-by-9 array of Xs.

How many Xs in all? ____

Write a number model.

____

6. True or false? Circle one.

The line segment is 6.2 centimeters long.

true false

____

SRB 119–121

Date Time

Math Boxes 5.12

1. Solve.

16 + 9 = ______________

16 + 90 = ______________

16 + 900 = ______________

16 + 9,000 = ______________

16 + 90,000 = ______________

2. Color 0.08 of the grid.

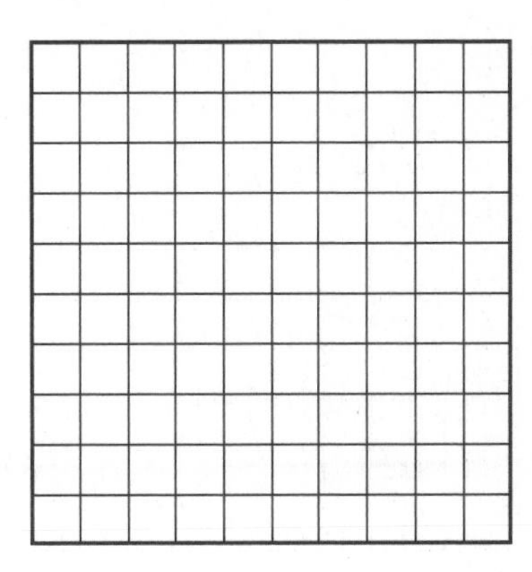

3. Find the differences between these high and low temperatures.

	High	Low	Difference
Pittsburgh	92°F	66°F	
Tempe	102°F	88°F	
Detroit	29°F	17°F	
Charlotte	37°F	23°F	

4. True or false? Circle one.

The line segment is 4.6 centimeters long.

true false

5. Add. Look for easy combinations.

25 + 13 + 5 = ______

19 + 11 + 23 = ______

33 + 14 + 27 = ______

6. Write the number that has

6 in the ones place

4 in the tenths place

3 in the hundredths place

2 in the thousandths place

____ • ____ ____ ____

Date Time

Math Boxes 5.13

1. Draw line segments *AB* and *CD*.

A • • *B*

C •

• *D*

SRB 88

2. Complete.

A triangle has _____ sides and _____ angles.

A quadrangle has _____ sides and _____ angles.

3. Draw a quadrangle.

4. Draw a polygon.

5. Circle the shape that has line symmetry.

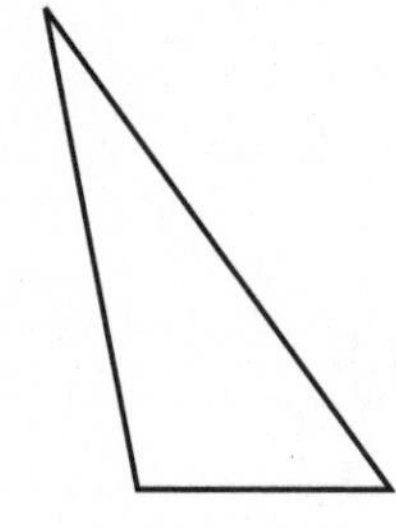

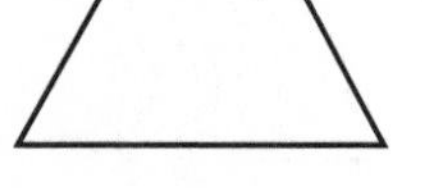

6. Circle the pictures that show 3-dimensional shapes.

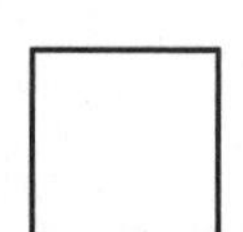

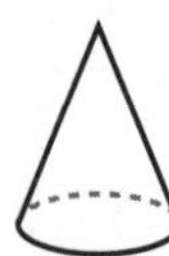

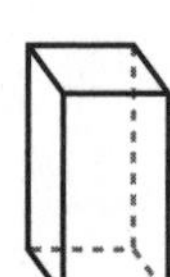

Date Time

Line Segments, Rays, and Lines

1. Write S next to each line segment. Write R next to each ray. Write L next to each line.

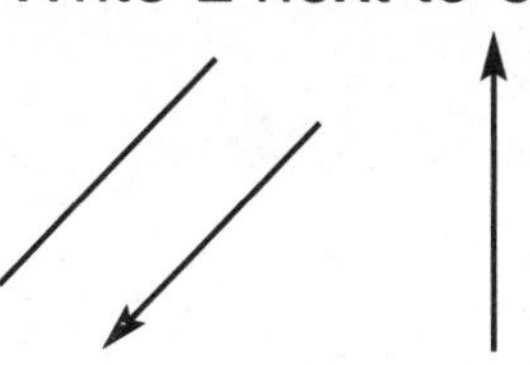

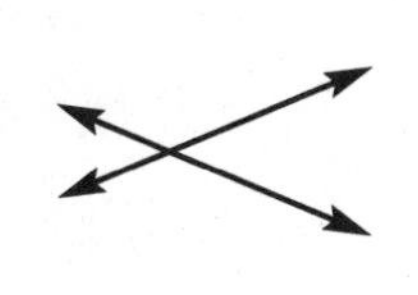

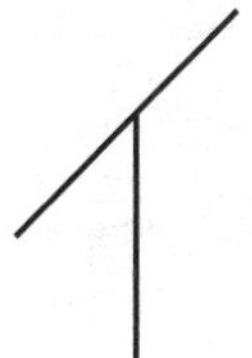

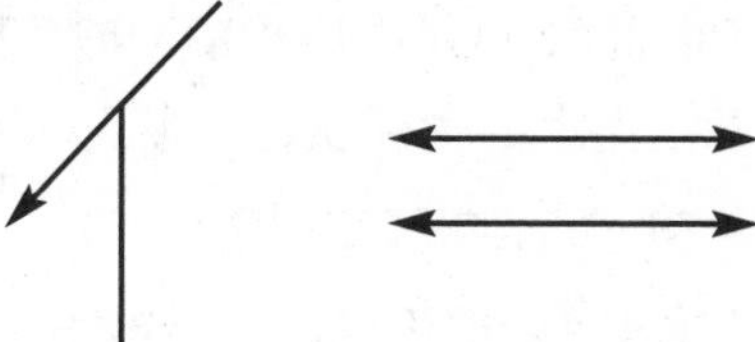

Points D, T, Q, and M are marked. Use a straightedge to draw the following.

2. Draw $\overline{QT}$. Draw $\overrightarrow{DT}$. Draw $\overleftrightarrow{MQ}$.

D • • T

M • • Q

Draw a line segment between each pair of points. How many line segments did you draw?

Example

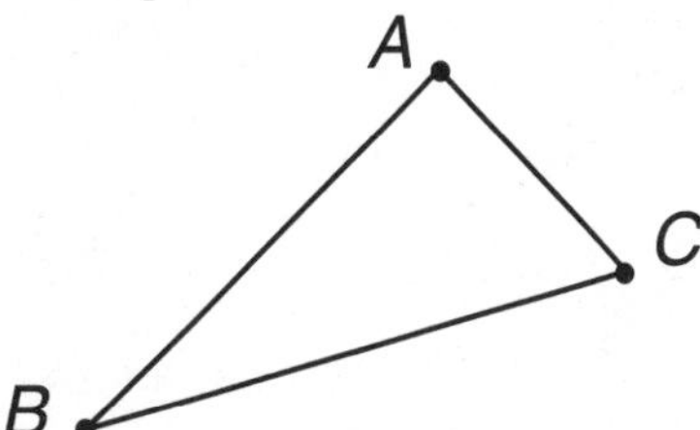

3 points

3 line segments

3. P • A • • L U •

4 points

_____ line segments

4. R • O • • E S • • I

5 points

_____ line segments

Math Boxes 6.1

1. Use the "about 3 times" circle rule to complete the table below:
For any circle, the circumference is about 3 times the diameter.

Unit
centimeters

diameter	circumference
	12
	18
	27

SRB 134 135

2. In the number 2.673

the 6 means 6 tenths

the 3 means ______

the 7 means ______

the 2 means ______

3. Fill in the unit box.
Then divide.

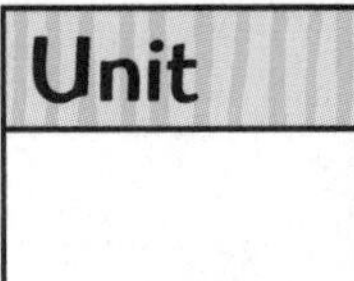
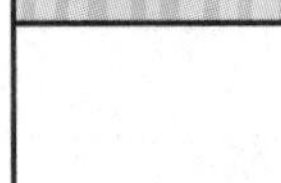

$30 \div 6 =$ ______

$12 \div 4 =$ ______

$20 \div 5 =$ ______

______ $= 14 \div 7$

______ $= 9 \div 3$

4. Write 4 division names for 6.

SRB 14 15 46 47

5. Write <, >, or =.

0.65 ______ 0.56

0.07 ______ 0.7

0.098 ______ 0.102

73.4 ______ 75.2

6. Solve.

$15 - 9 =$ ______

$25 - 9 =$ ______

$55 - 9 =$ ______

$85 - 9 =$ ______

$105 - 9 =$ ______

Date Time

Geometry Hunt

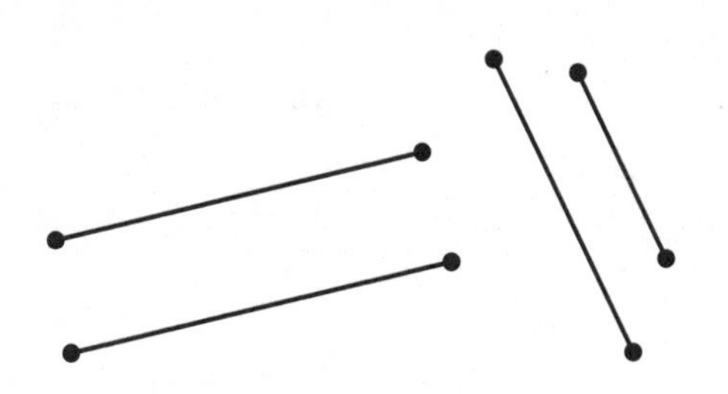
parallel line segments

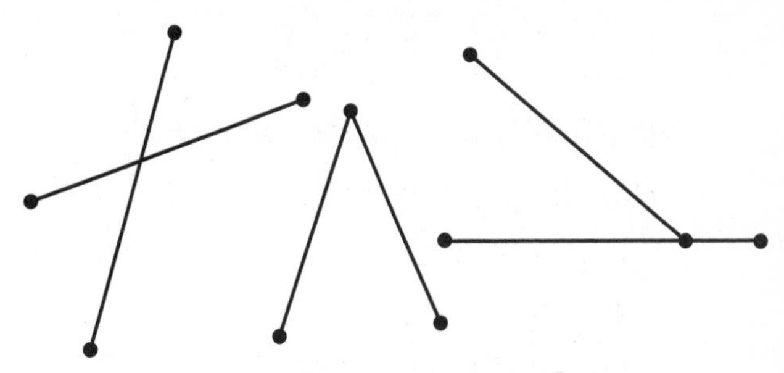
intersecting line segments

Part 1

Look for things in the classroom or hallway that are parallel. Look for things that intersect. List these things below or draw a few of each of them on another sheet of paper.

Parallel

____

____

____

Intersecting

____

____

____

Part 2

Look for things in the classroom or hallway that have one or more right angles. List these things below or draw a few of them on another sheet of paper.

____

____

____

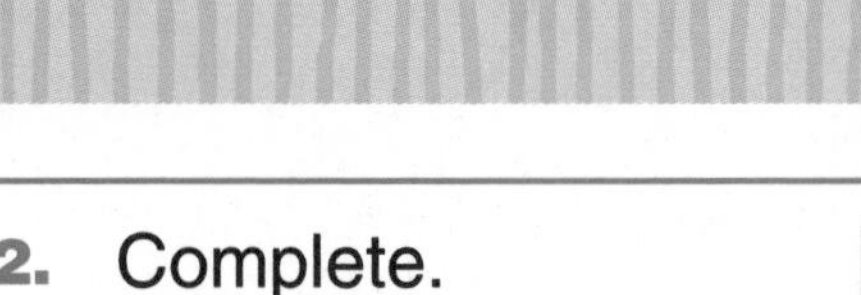

Math Boxes 6.2

1. Draw a ray, $\overrightarrow{AB}$. Draw a line segment, $\overline{CD}$. Draw a line, $\overleftrightarrow{EF}$.

• A • B

• C • D

• E • F

2. Complete.

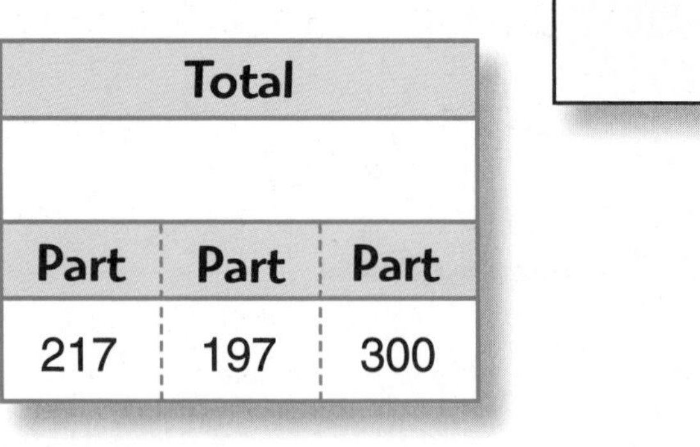

Unit

Total		
Part	Part	Part
217	197	300

Number model: ______________________

3. Complete the Fact Triangle. Write the fact family.

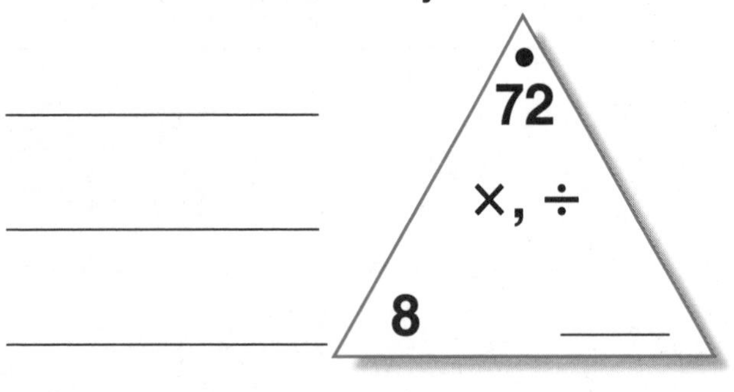

____

____

____

____

4. Complete.

in	out
16	
	240
225	
133	
	1,000

in → **Rule**: double → out

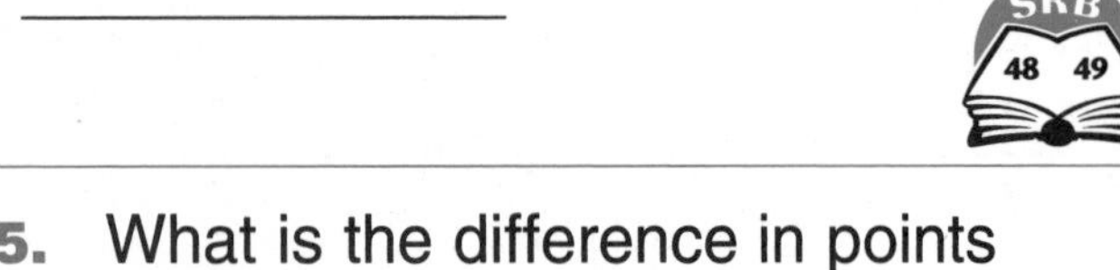

5. What is the difference in points between Players B and C?

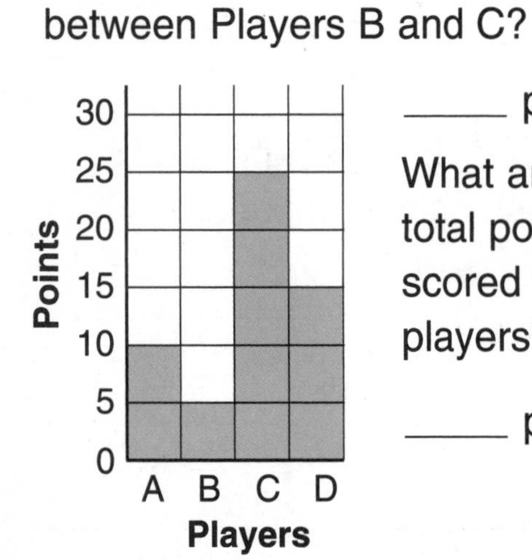

_____ points

What are the total points scored for all players?

_____ points

6. Write equivalent lengths.

$\frac{1}{3}$ yd = _____ ft

18 in. = _____ yd

50 mm = _____ cm

0.6 m = _____ cm

Date Time

Turns

Use your connected straws to show each turn.
Draw a picture of what you did.
Draw a curved arrow to show the direction of the turn.

Example

Example	1.	2.
right $\frac{1}{4}$ turn (clockwise) 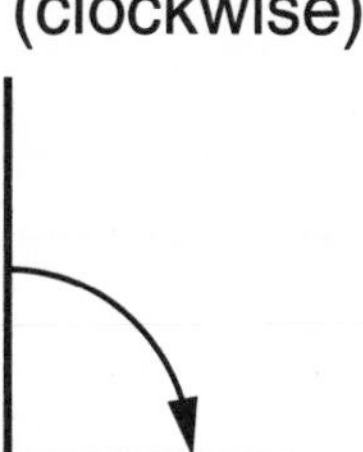	right $\frac{1}{2}$ turn (clockwise)	left $\frac{1}{4}$ turn (counterclockwise)
3. left $\frac{3}{4}$ turn (counterclockwise)	**4.** right $\frac{3}{4}$ turn (clockwise)	**5.** left $\frac{1}{2}$ turn (counterclockwise)

Date Time

Math Boxes 6.3

1. Circle the pair of lines that are parallel.

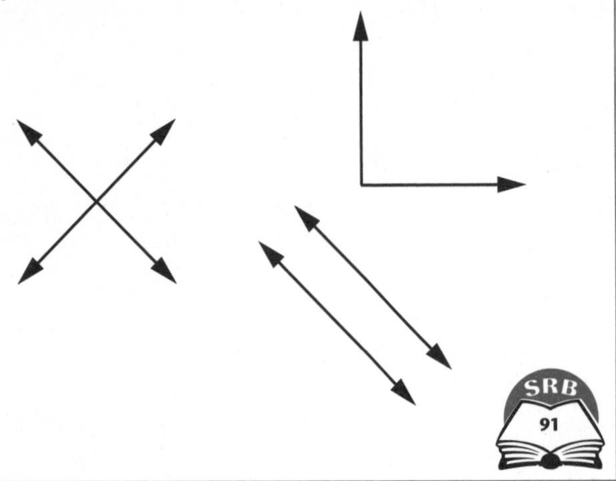

2. Annette had \$16.75 in her purse. She spent \$8.66 at the store. How much money does she have left?

____

3. Write <, >, or =.

4×7 _____ 5×6

7×5 _____ 6×3

4×6 _____ 5×5

5×7 _____ 4×9

4. Solve.

$4{,}695 + 1{,}013 =$ _______

$5{,}692 - 3{,}688 =$ _______

_______ $= 10{,}000 + 695$

5. Complete.

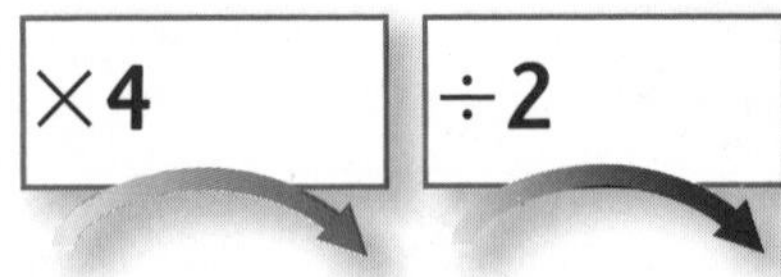

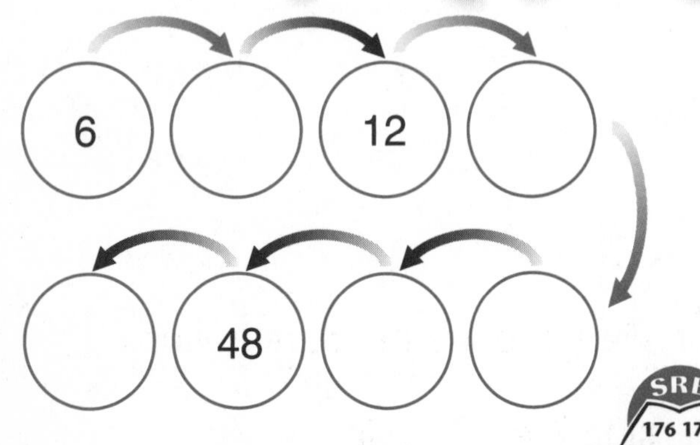

6. Rectangle *ABCD* is a(n)

_____ by _____ rectangle.

The area of rectangle *ABCD*:

_____ × _____ = _____ square units.

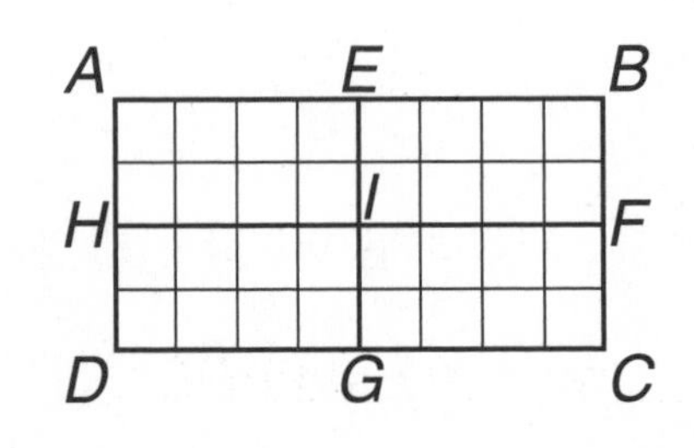

 Use with Lesson 6.3.

Date Time

Triangle Explorations

Part 1

Follow these steps:

1. Mark three points on the circle.
2. Label them *A, B,* and *C.*
3. Use a straightedge to connect each pair of points with a line segment.
4. What figure have you drawn?

____

Part 2

Write all six 3-letter names that are possible for your triangle. The first letter of each name is given below.

*A*______ *A*______ *B*______ *B*______ *C*______ *C*______

Part 3

Work with a group.

Make triangles with straws and twist-ties. Make at least one of each of the following kinds of triangles:

- all 3 sides the same length
- only 2 sides the same length
- no sides the same length
- 1 right angle
- 1 angle larger than a right angle
- all 3 angles smaller than a right angle

Part 4

Measure each side of the triangle you drew in Part 1 to the nearest $\frac{1}{4}$ inch.

side *AB* _____ in. side *BC* _____ in. side *CA* _____ in.

Date Time

Math Boxes 6.4

1. Draw a ray, $\overrightarrow{AT}$. Draw a line segment, $\overline{BY}$. Draw a line, $\overleftrightarrow{ME}$.

• A • T

• B • Y

• M • E

SRB 88 89

2. In the number 34.972

the 9 means 0.9

the 7 means ______

the 3 means ______

the 4 means ______

the 2 means ______

3. The turn of the angle is

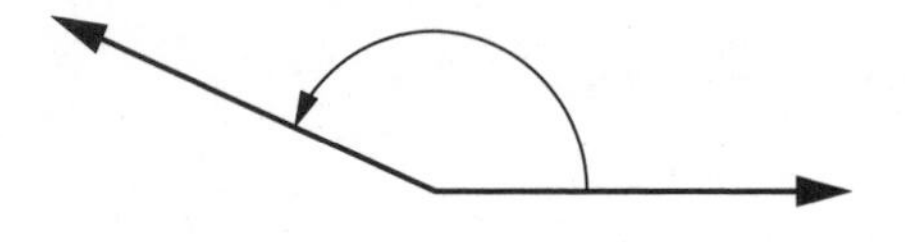

0 less than a $\frac{1}{2}$ turn.

0 less than a $\frac{1}{4}$ turn.

0 greater than a $\frac{1}{2}$ turn.

0 a full turn.

4. Double these numbers:

6 → ____

8 → ____

12 → ____

Triple these numbers:

4 → ____

6 → ____

11 → ____

5. Write the time in hours and minutes.

half-past 6 ____:____

quarter-past 9 ____:____

quarter to 12 ____:____

10 minutes to 10 ____:____

6. What temperature is it? ________

°F
50
40
30
20
10

How many degrees difference is there between 90°F and the above temperature?

____

Date Time

Quadrangle Explorations

Part 1 Follow these steps:

1. Mark four points on the circle.
2. Label the points *A, B, C,* and *D.*
3. Use a straightedge to connect pairs of points to form a quadrangle.

Part 2 Write all eight 4-letter names that are possible for your quadrangle. The first letter of each name is given below.

A ________ *A* ________ *B* ________ *B* ________

C ________ *C* ________ *D* ________ *D* ________

Part 3 Work with a group.

Make quadrangles with straws and twist-ties. Make at least one of each of the following kinds of quadrangles:

- all 4 sides equal in length
- 2 pairs of equal-length sides, but opposite sides not equal length
- 2 pairs of equal-length opposite sides
- only 2 parallel opposite sides, each a different length
- only 1 pair of equal-length opposite sides

Part 4 Measure each side of the quadrangle you drew in Part 1 to the nearest $\frac{1}{4}$ inch.

side *AB* _____ in. side *BC* _____ in. side *CD* _____ in. side *DA* _____ in.

Estimate: The perimeter of my quadrangle is about _____ inches.

Date Time

Math Boxes 6.5

1. Circle the lines that intersect.

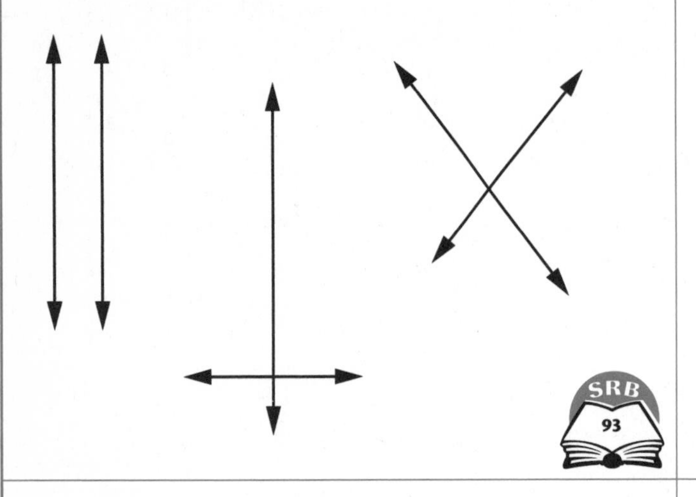

SRB 93

2. Circle the right angle.

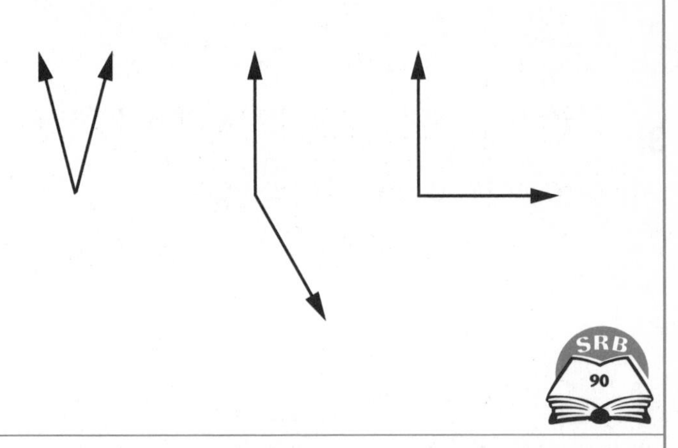

SRB 90

3. Write $<$, $>$, or $=$.

10×0 _____ 429×0

7×6 _____ 6×6

5×4 _____ 4×5

1×18 _____ 4×4

4. Measure each side of the triangle to the nearest centimeter.

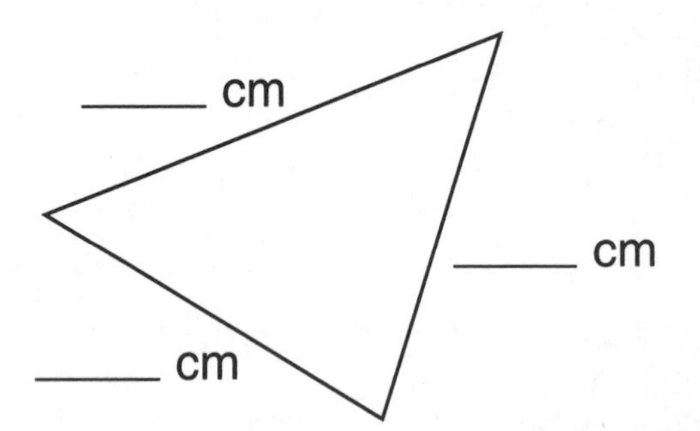

Perimeter = _____ cm

5. Complete the bar graph.

Lily ran 4 miles.

Meg ran 3 miles.

Rita ran 6 miles.

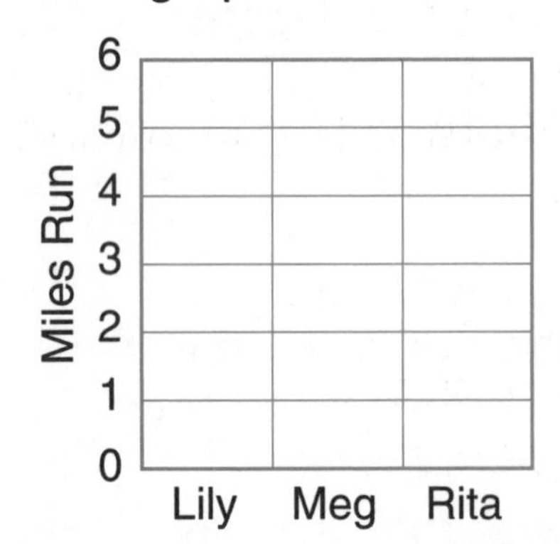

Median miles run: _____

6. Which is more?

1.36 or 1.6 _______

0.4 or 0.372 _______

0.69 or 0.6 _______

0.7 or 0.09 _______

Polygon Explorations

Part 1 Follow these steps:

1. Mark 5 points on the circle.
2. Label the points *A, B, C, D,* and *E*.
3. Use a straightedge to connect pairs of points to form a polygon.
4. What kind of polygon is it? ______________
5. Write 4 or more possible names for your polygon.

____________ ____________ ____________

____________ ____________ ____________

Part 2 Work with a group.

Make polygons with straws and twist-ties. Your teacher will tell you how many sides your polygons should have.

Make at least one of each of the following kinds of polygons:

- all sides equal in length, and all angles equal in size (the amount of turn between sides)
- all sides equal in length, but not all angles equal in size
- *any* polygon having the assigned number of sides

Date Time

Polygon Explorations (cont.)

Part 3 A **regular polygon** is a polygon in which all the sides are equal and all the angles are equal.

Below, trace the smaller of each kind of *regular* polygon from your Pattern-Block Template.

Below, trace all the polygons from your Pattern-Block Template that are *not* regular polygons.

Part 4 Measure each side of the polygon you drew in Part 1 to the nearest $\frac{1}{2}$ centimeter.

side *AB* __________ cm

side *BC* __________ cm

side *CD* __________ cm

side *DE* __________ cm

side *EA* __________ cm

Estimate: The perimeter of my polygon is about __________ cm.

Date Time

Math Boxes 6.6

1. Draw a ray, $\overrightarrow{SO}$. Draw a line segment, $\overline{LA}$. Draw a line, $\overleftrightarrow{TI}$.

SRB 88 89

2. Fill in the missing numbers. Add going across and subtract going down.

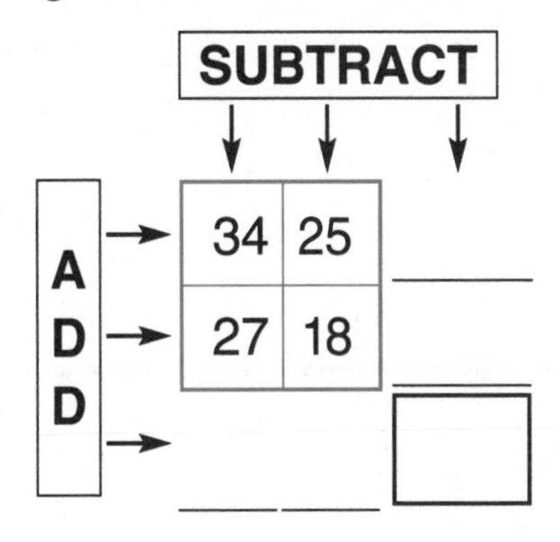

3. The turn of the angle is

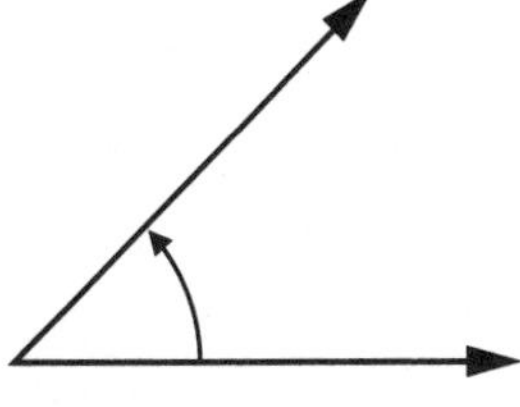

0 greater than a $\frac{1}{4}$ turn.

0 less than a $\frac{1}{4}$ turn.

0 greater than a $\frac{1}{2}$ turn.

0 a full turn.

4. 9 boxes of muffins. 6 muffins per box. How many muffins in all?

______ muffins

Write a number model:

____ × ____ = ____

5. Draw a shape with 4 sides that are all equal in length.

This shape is a ______________.

6. Complete the Fact Triangle. Write the fact family.

____

____

____

____

21

×, ÷

3 ____

SRB 48 49

Date Time

Drawing Angles

Draw each angle as directed by your teacher.
Record the direction of each turn with a curved arrow.

Part 1

A •

B •

C •

Part 2

R •

S •

T •

Use with Lesson 6.7.

Math Boxes 6.7

1. Label all the points of these intersecting lines. Name the 2 lines.

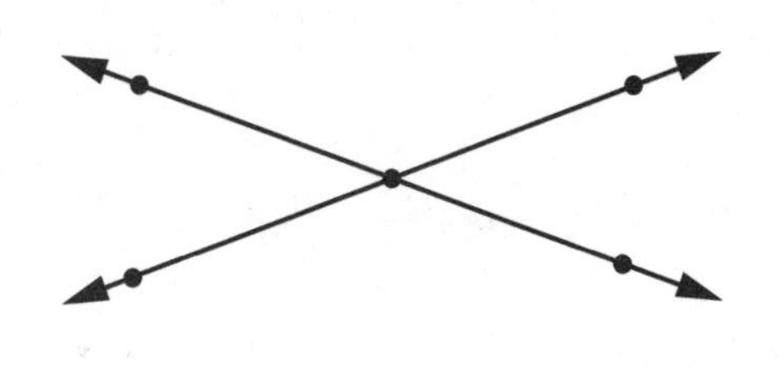

____

2. Circle the regular polygons.

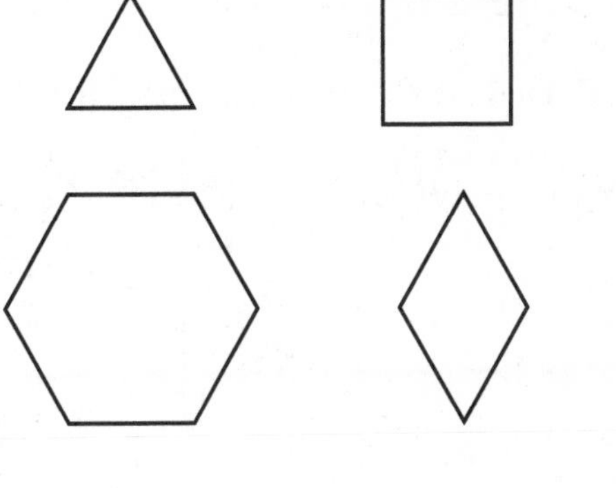

3. Fill in the unit box. Then divide.

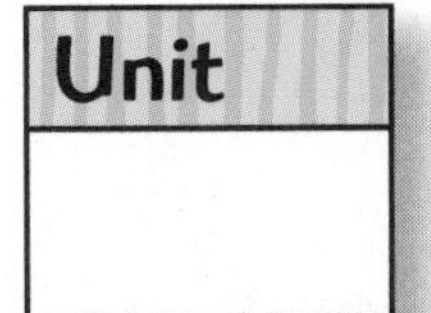

25 ÷ 5 = _____

18 ÷ 3 = _____

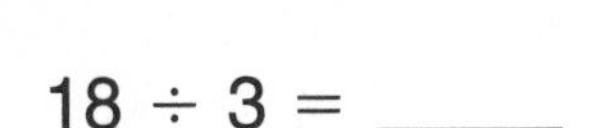

_____ = 30 ÷ 5

28 ÷ 7 = _____

_____ = 24 ÷ 4

4. Circle the right angle.

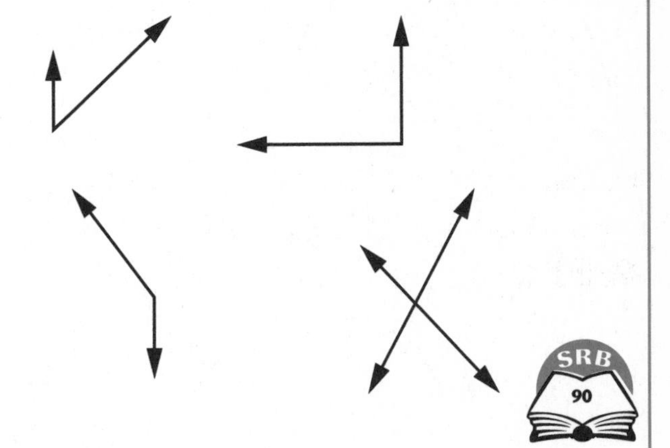

5. Measure each side of the triangle to the nearest centimeter.

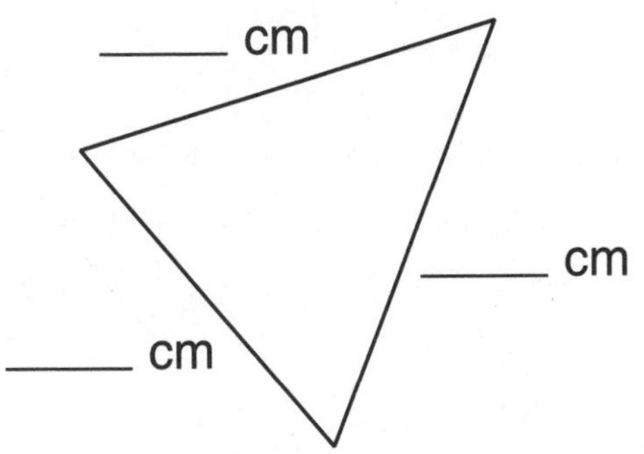

Perimeter = _____ cm

6. In the number 4.908

the 4 means *4 ones*

the 0 means ______________

the 9 means ______________

the 8 means ______________

Date Time

Marking Angle Measures

Connect 2 straws with a twist-tie. Bend the twist-tie at the connection.

Place the straws on the circle.

- Place the bend on the center of the circle.
- Place both straws pointing to 0°.

Keep one straw pointing to 0°. Move the other straw to form angles.

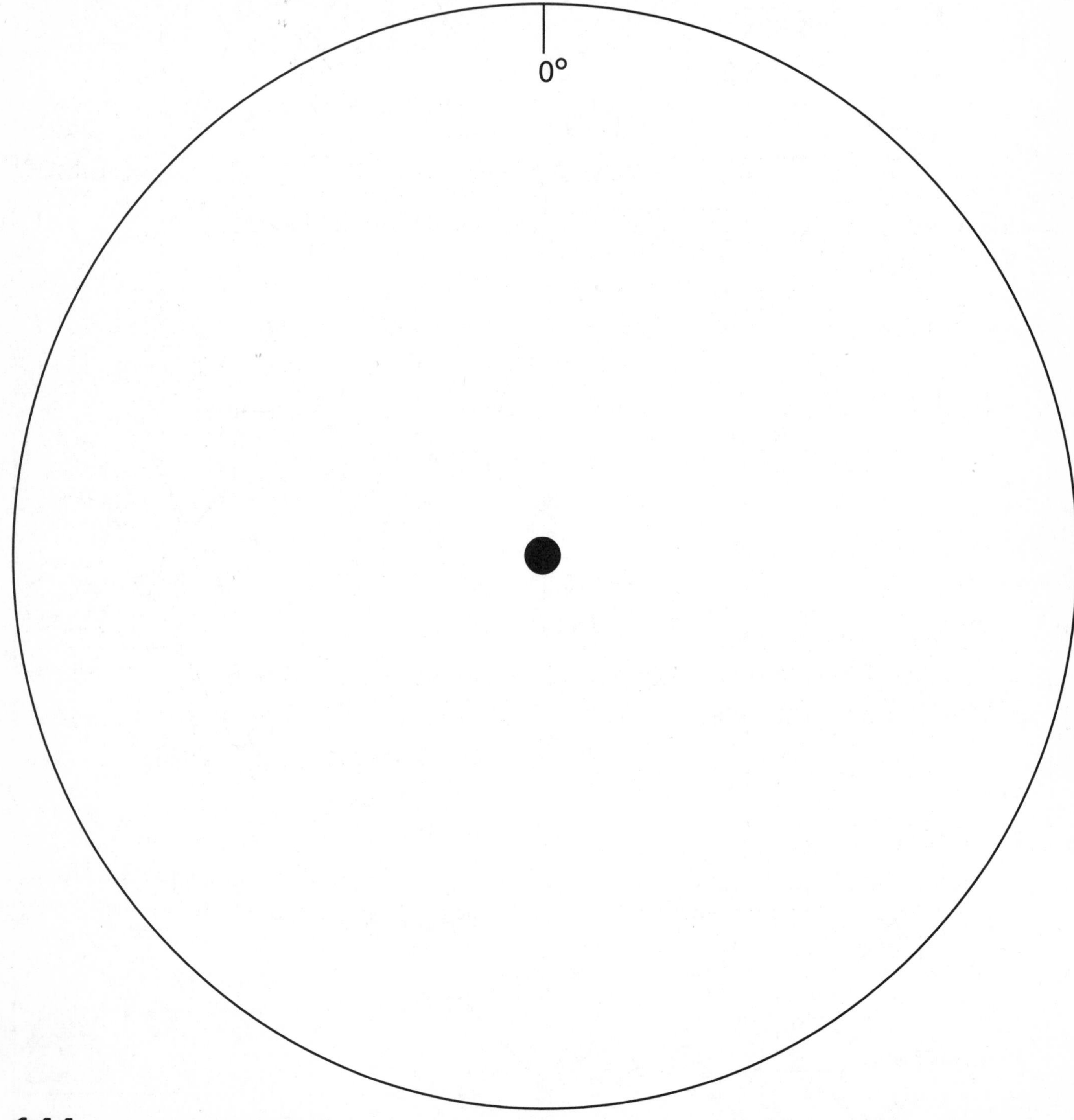

Date Time

Measuring Angles

Use your angle measurer to measure the angles on this page.
Record your measurements in the table.

Angle	Measurement
A	about ______°
B	about ______°
C	between ______° and ______°
D	about ______°
E	about ______°
F	about ______°

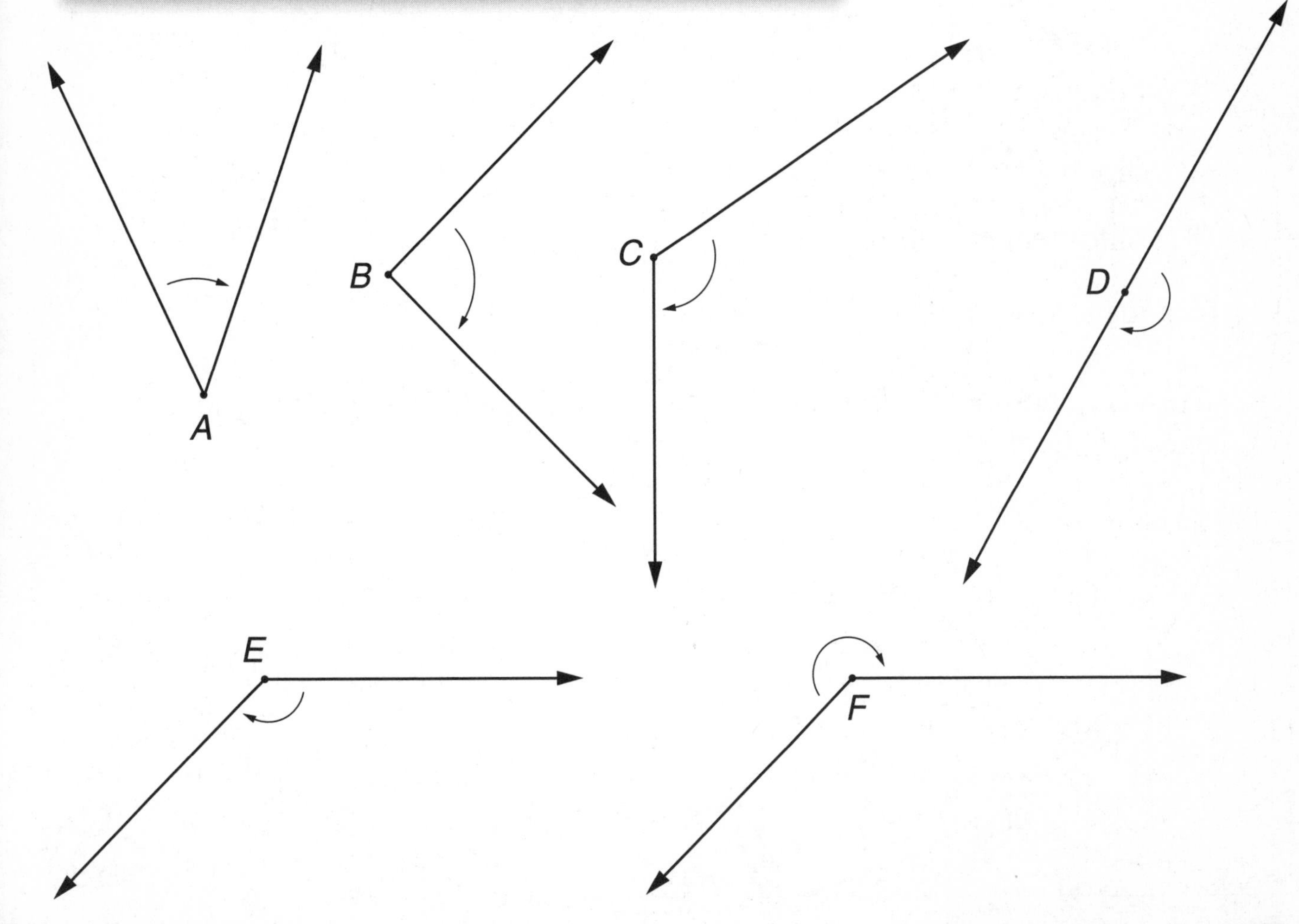

Date Time

Math Boxes 6.8

1. Draw a ray, $\overrightarrow{DO}$. Draw a line segment, $\overline{RE}$. Draw a line, $\overleftrightarrow{MI}$.

SRB 88 89

2. Complete the equal-sharing story.

14 ________ are shared equally by ____ girls.

How many ________ does each ________ get? ____

How many ________ are left over? ____

3. The turn of the angle is

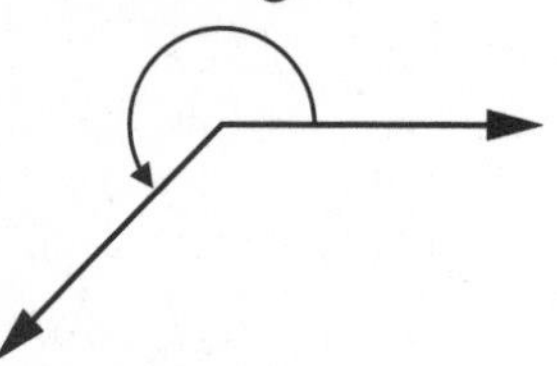

0 greater than a $\frac{3}{4}$ turn.

0 less than a $\frac{1}{4}$ turn.

0 greater than a $\frac{1}{2}$ turn.

0 a full turn.

4. Double these numbers:

15 → ______

80 → ______

200 → ______

Triple these numbers:

10 → ______

25 → ______

300 → ______

5. Draw a quadrangle with exactly one right angle.

SRB 98 99

6. Complete the number-grid puzzle.

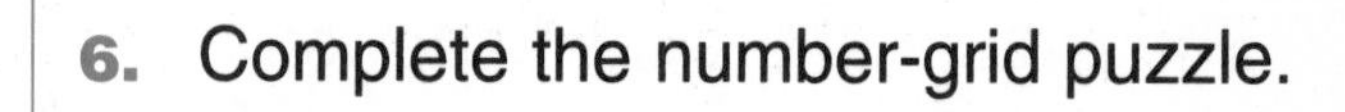

2,014

SRB 7 8

Use with Lesson 6.8.

Date Time

Symmetric Shapes

Each picture below shows one half of a letter. The dashed line is the line of symmetry. Guess what the letter is. Then draw the other half of the letter.

1.

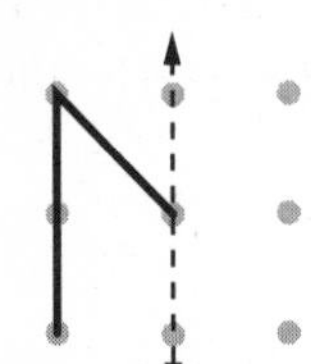

2.

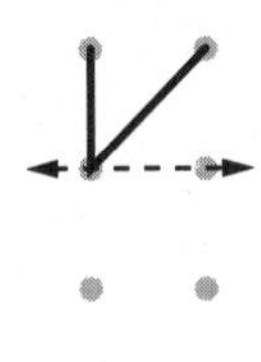

3.

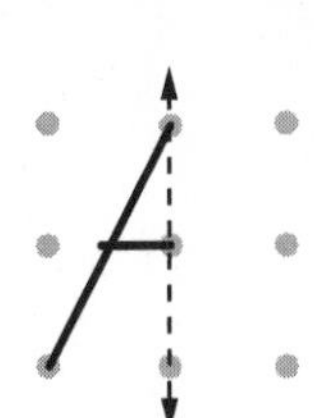

4.

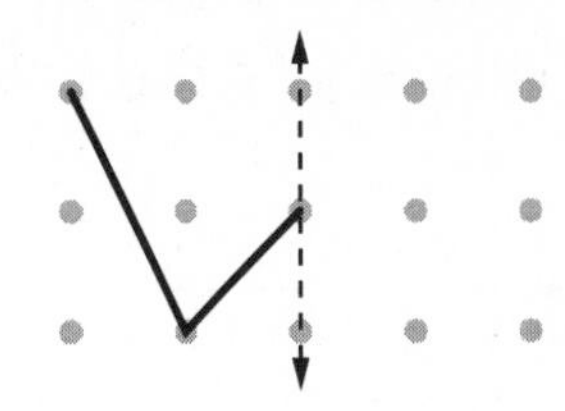

Draw the other half of each symmetric shape below.

5.

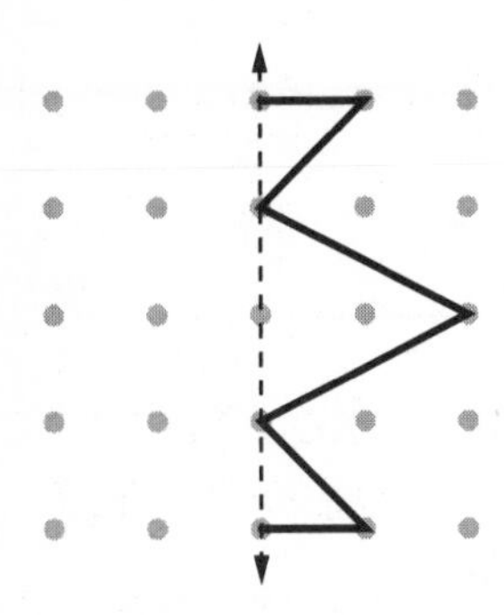

6.

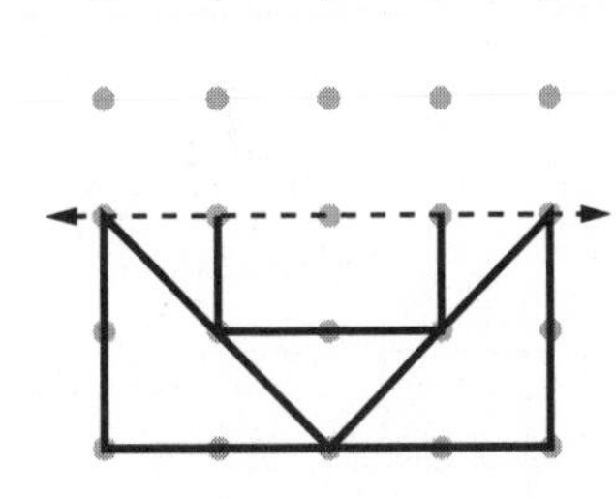

7.

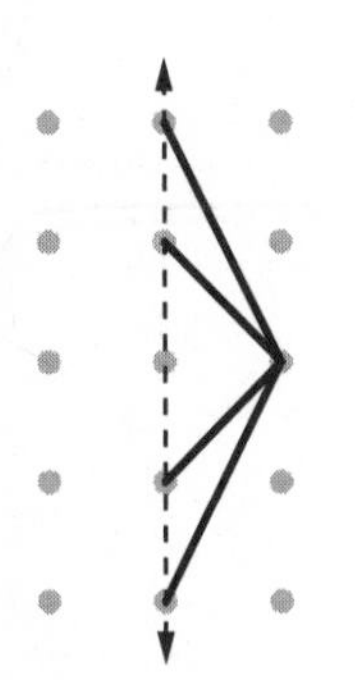

8.

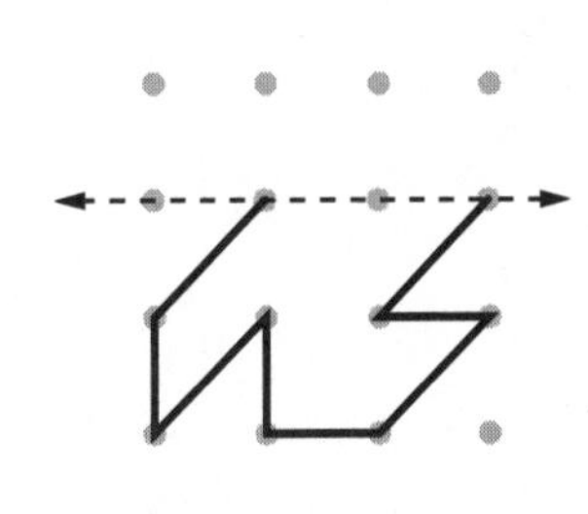

9. The picture at the right shows one-fourth of a symmetric shape. There are two lines of symmetry. Draw the mirror image for each line of symmetry.

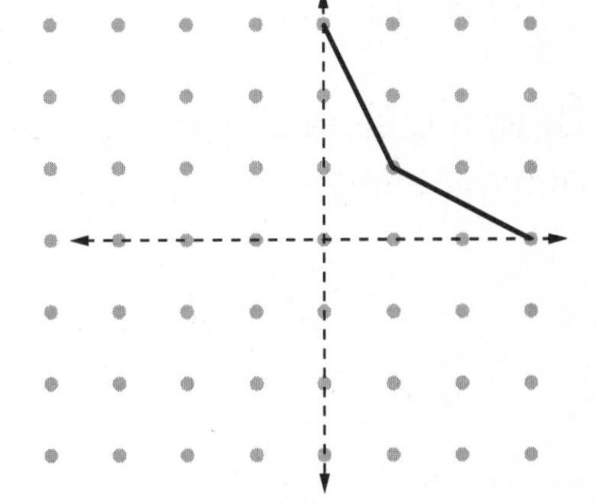

Challenge

10. There are 2 more lines of symmetry in Problem 9. Draw them.

Date Time

Math Boxes 6.9

1. Draw a line segment, $\overline{DI}$, parallel to the line, $\overleftrightarrow{PO}$. Draw a ray, $\overrightarrow{LA}$, that intersects the line, $\overleftrightarrow{TW}$.

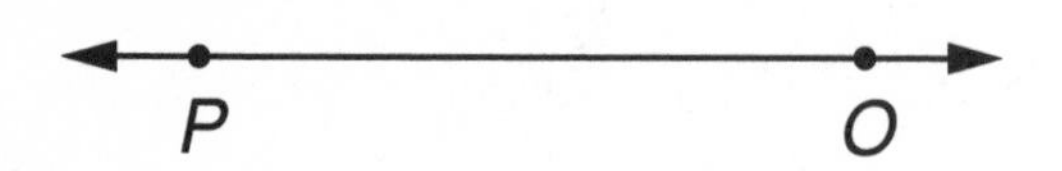

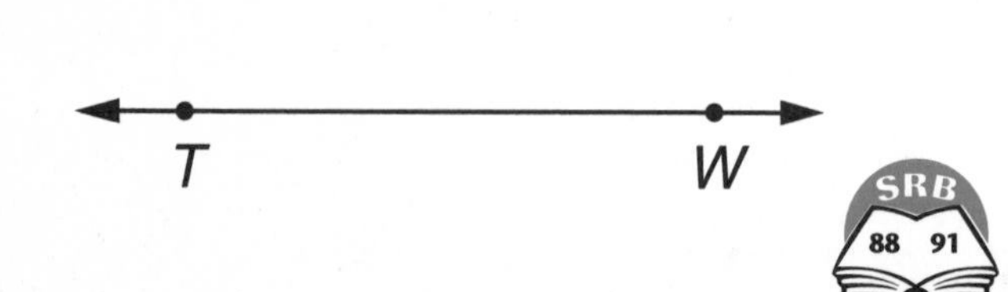

2. Describe a regular polygon.

____

____

____

____

____

____

3. Fill in the unit box. Then multiply.

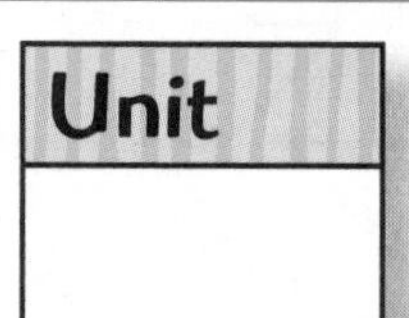

$4 \times 5 =$ _____

$7 \times 3 =$ _____

_____ $= 4 \times 4$

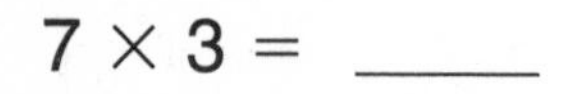

_____ $= 5 \times 3$

_____ $= 7 \times 5$

4. The degree measure of the angle is

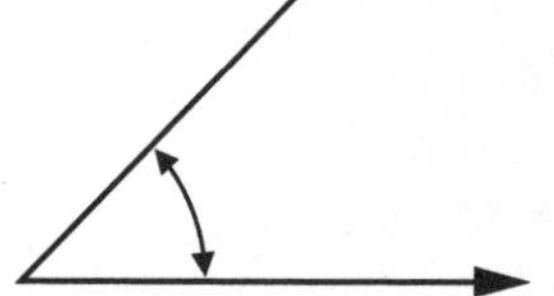

0 more than 90°.

0 less than 90°.

0 more than 180°.

0 120°.

5. Measure each side of the quadrangle to the nearest half-centimeter.

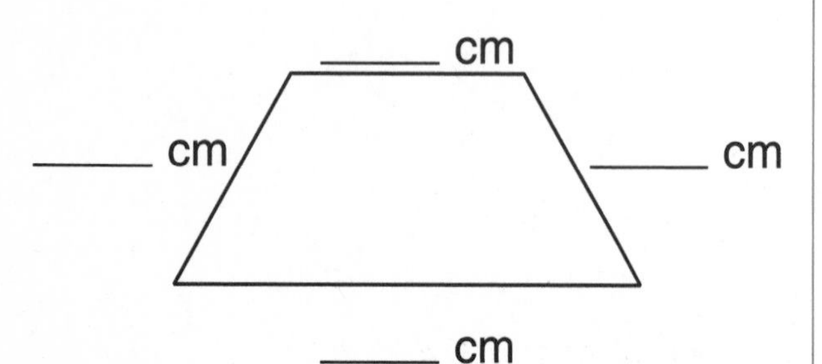

Another name for this quadrangle is a ______________.

SRB 99 119–121

6. Circle the right angle.

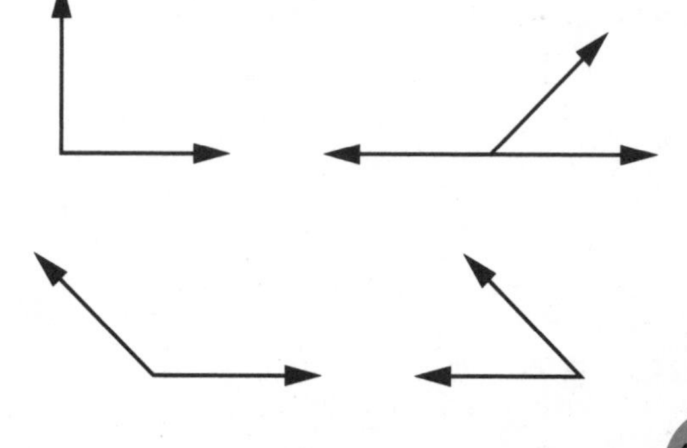

SRB 90

Date Time

Math Boxes 6.10

1. These letters are *Symmets:*

H, T, M, A

This letters are not *Symmets:*

F, J, R, S

Write other letters that are *Symmets:*

____

2. Write the number that has

7 in the thousandths place

5 in the ones place

1 in the tenths place

3 in the hundredths place

__ . __ __ __

3. The turn of the angle is

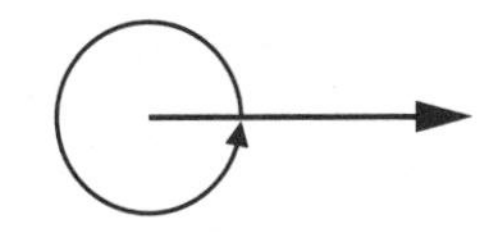

0 a $\frac{1}{4}$ turn.

0 less than a $\frac{1}{4}$ turn.

0 less than a $\frac{1}{2}$ turn.

0 a full turn.

4. Read the graph.

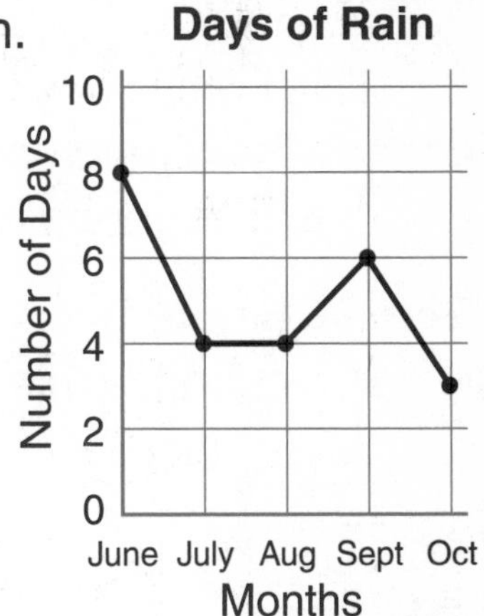

Which month had the most days of rain?

____

What is the median number of days of rain? ______

5. Draw a quadrangle with exactly one pair of parallel sides.

This shape is called a

________________.

6. Draw a 4-by-8 array of Xs.

How many Xs in all? ______

Write a number model.

____

SRB 63 64

Date Time

Base-10 Block Decimal Designs

Materials ❑ base-10 blocks (cubes, longs, and flats)
❑ 10-by-10 grids (*Math Journal 1,* p. 151)
❑ crayons or colored pencils

Think of the *flat* as a unit, or ONE. Remind yourself of the answers to the following questions:

- How many cubes would you need to cover the whole flat?
- How much of the flat is covered by 1 cube? By 1 long?

Follow these steps:

Step 1 Make a design by putting some cubes on a flat.

Step 2 Copy your design in color onto one of the grids on journal page 151.

Step 3 How much of the flat is covered by the cubes in your design? To help you find out, exchange as many cubes as you can for longs.

Step 4 Figure out which decimal tells how much of the flat is covered by cubes. Write the decimal under the grid that has your design on it.

Example

Steps 1 and 2: Make a design on a flat with cubes. Copy the design onto a grid.

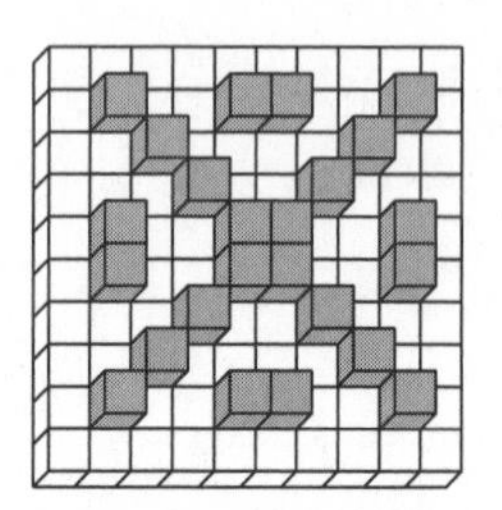

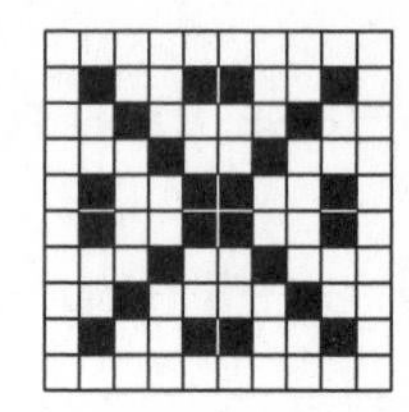

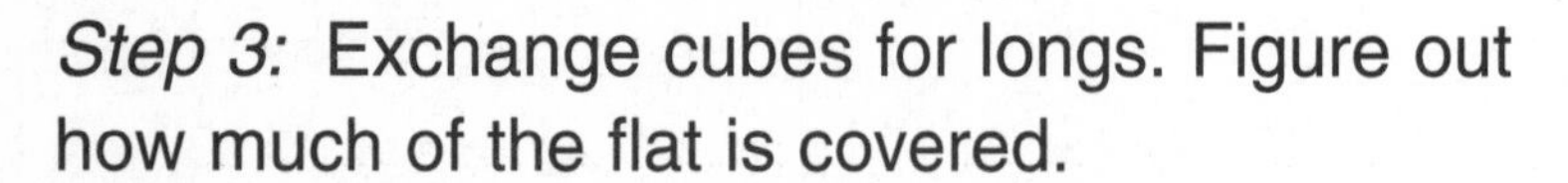

Step 3: Exchange cubes for longs. Figure out how much of the flat is covered.

Step 4: Write the decimal under the grid.

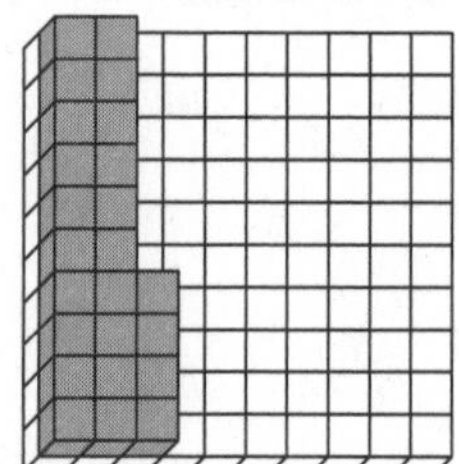

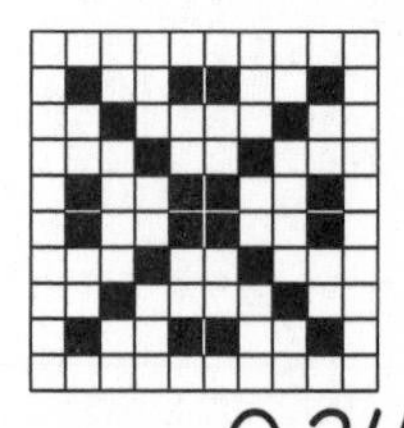

Decimal: 0.24

Make other designs with cubes on flats, and draw them on the grids. Write a decimal for each design.

Date Time

10 × 10 Grids

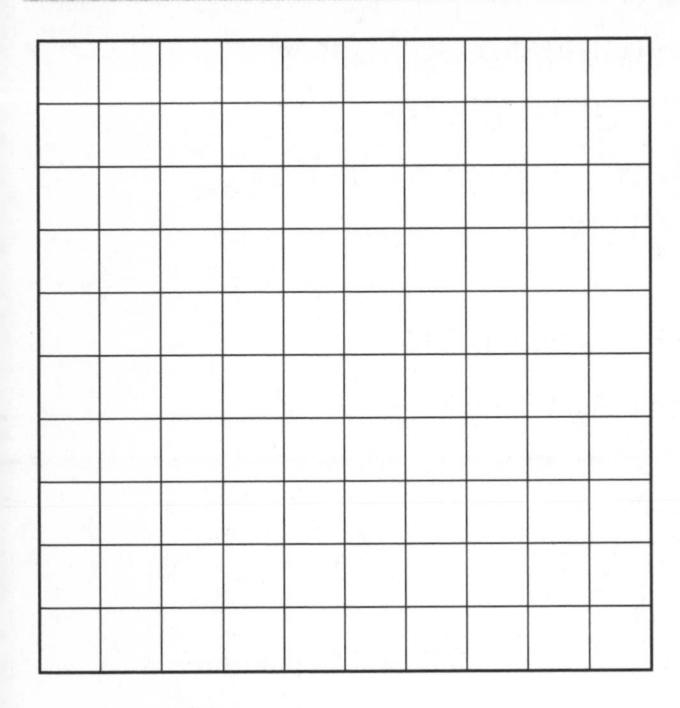

Decimal: ____________________

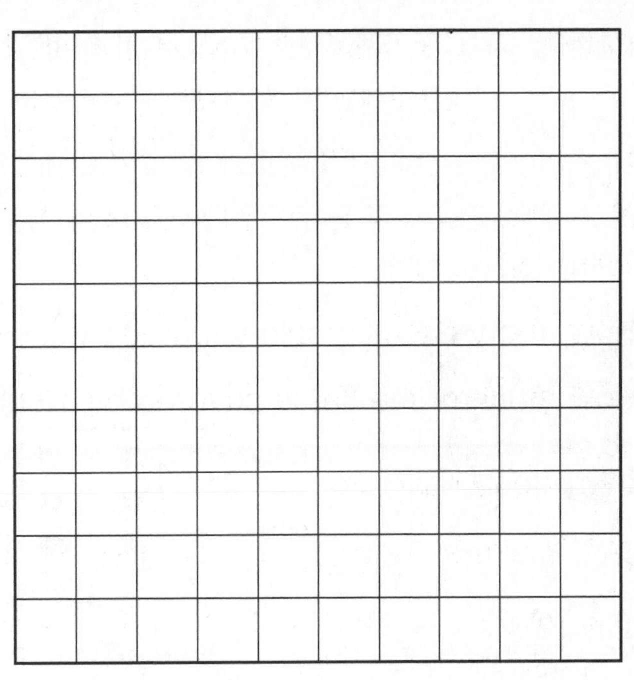

Decimal: ____________________

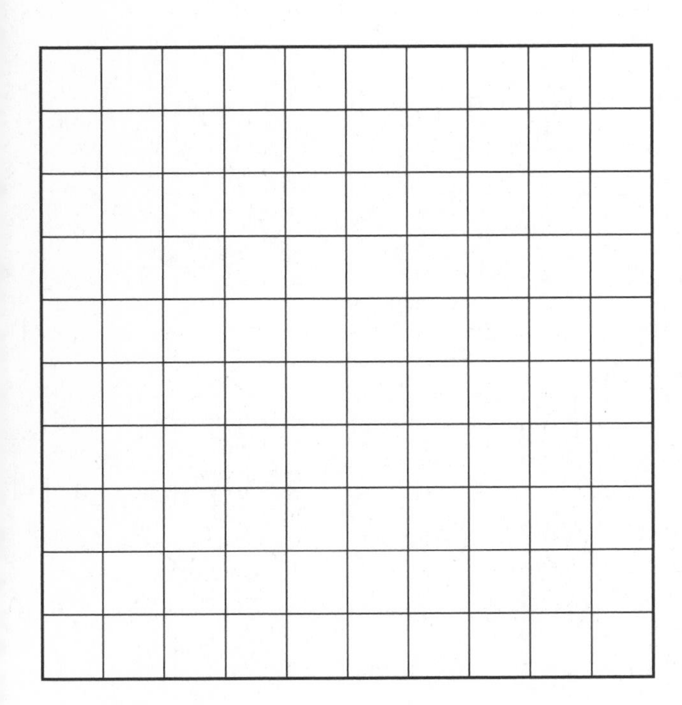

Decimal: ____________________

Decimal: ____________________

Date Time

Symmetry

If a shape can be folded in half so that the two halves match exactly, the shape is **symmetric.** We also say that the shape has **symmetry.**

The fold line is called the **line of symmetry.** Some symmetric shapes have just one line of symmetry. Others have more.

1 line of symmetry

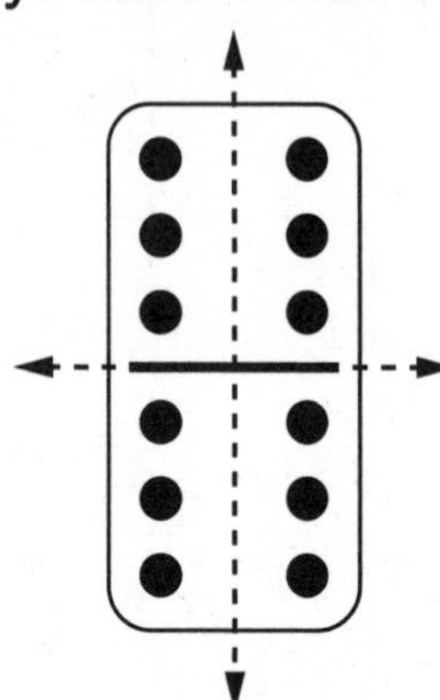

2 lines of symmetry

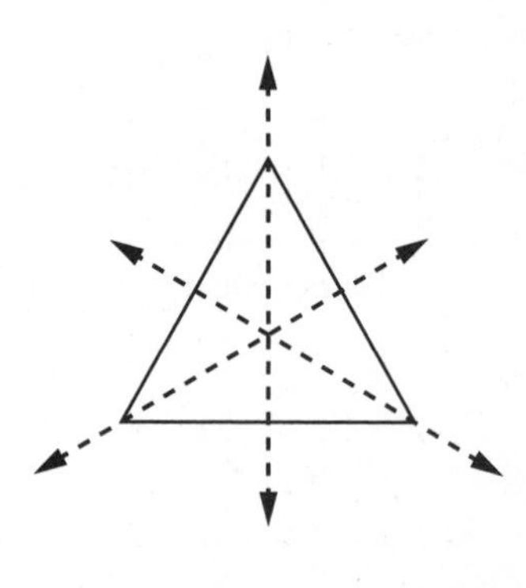

3 lines of symmetry

1. Which of the following shapes is **not** symmetric? ______

a.

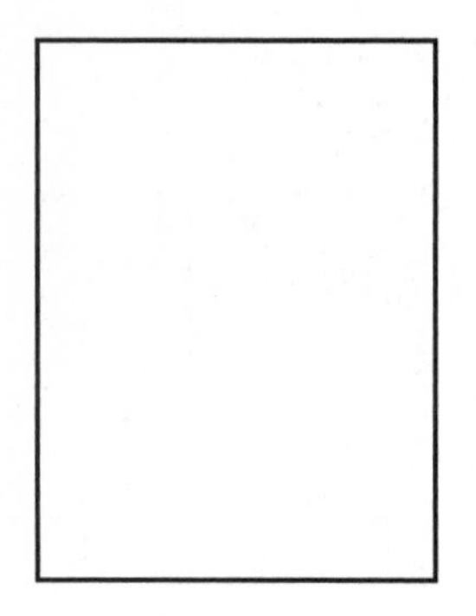

b.

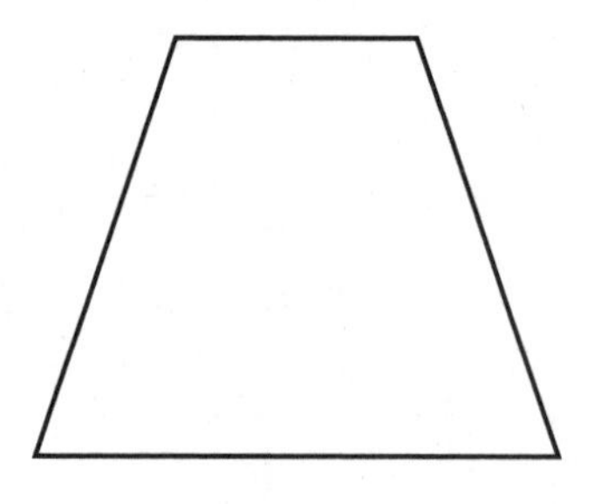

c.

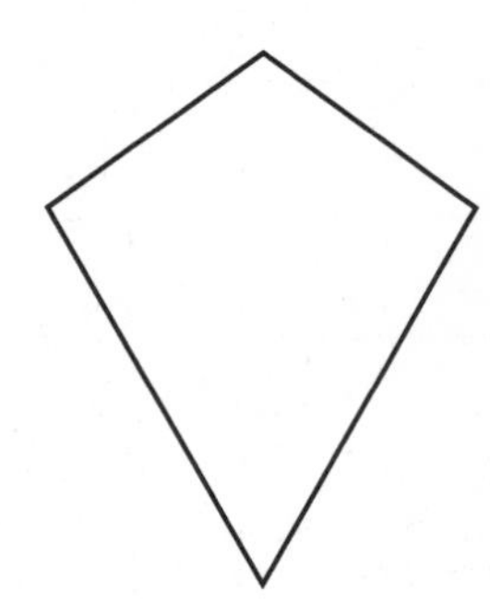

d.

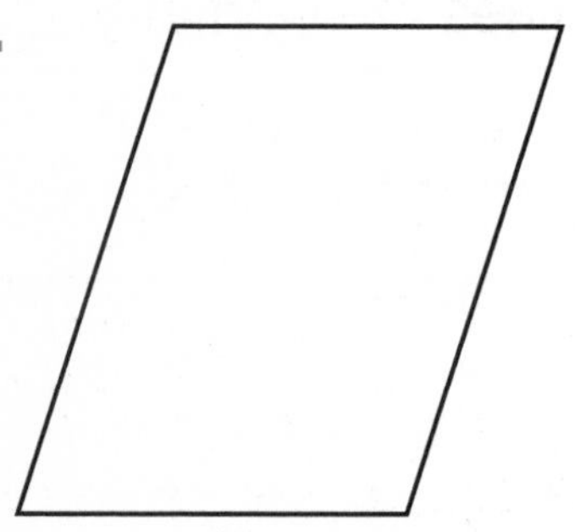

e.

f.

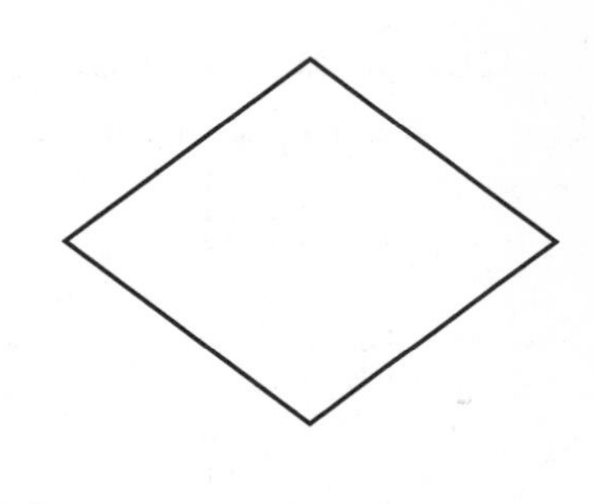

2. Draw all the lines of symmetry on the shapes that are symmetric.

Math Boxes 6.11

1. If a map scale shows that $\frac{1}{2}$ inch represents 50 miles, then

1 inch represents _______ miles

2 inches represents _______ miles

4 inches represents _______ miles

_______ inches represents 500 miles

SRB 164

2. Figure out this riddle:

I have four sides. My opposite sides are equal in length. One set of my sides is longer than the other set of my sides. What shape am I?

____

3. Fill in the unit box. Then divide.

Unit

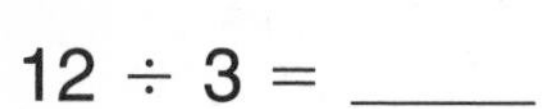
$12 \div 3 =$ _____

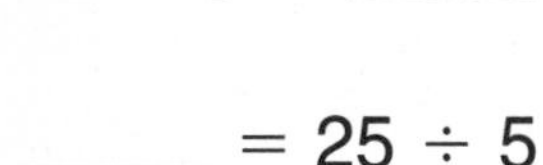
_____ $= 25 \div 5$

_____ $= 28 \div 4$

_____ $= 21 \div 7$

$24 \div 4 =$ _____

4. The degree measure of the angle is

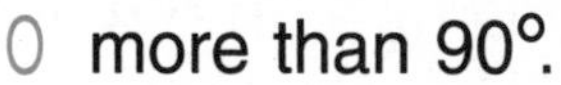

○ more than 90°.

○ less than 90°.

○ more than 180°.

○ 40°.

5. Measure the sides of the quadrangle to the nearest centimeter.

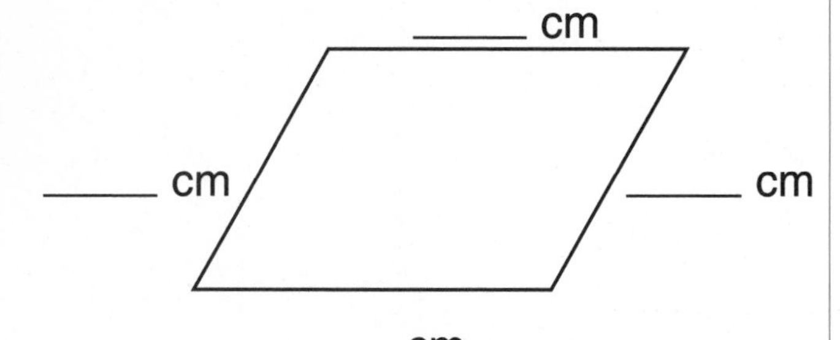

Another name for this quadrangle is

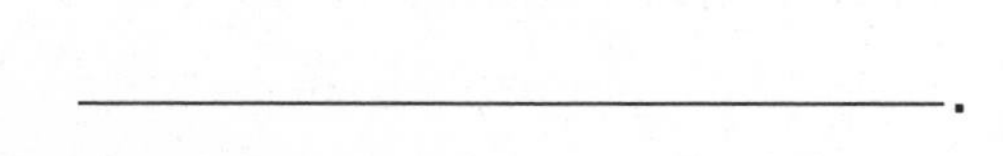
_____________________________.

6. Draw the lines of symmetry.

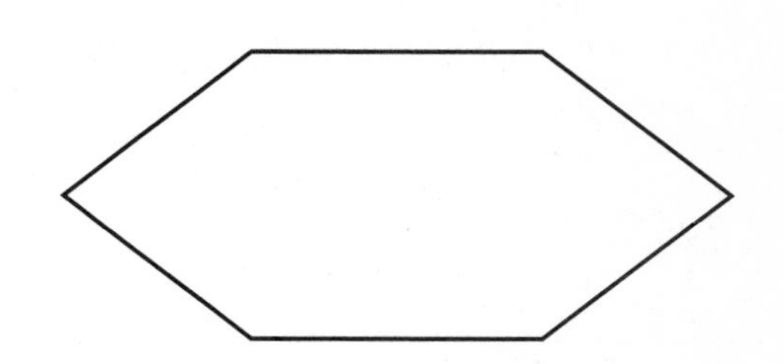

There are _____ lines of symmetry.

Date Time

Pattern-Block Prisms

Work in a group.

1. Each person chooses a different pattern-block shape.
2. Each person then stacks 3 or 4 of the shapes together. See below.
3. Each person makes a prism by using small pieces of tape to hold the blocks together.

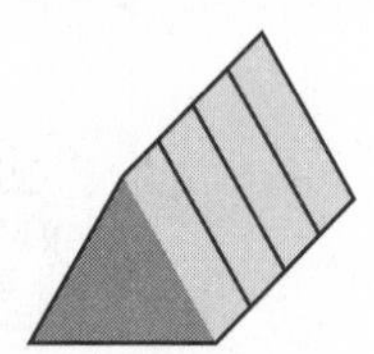

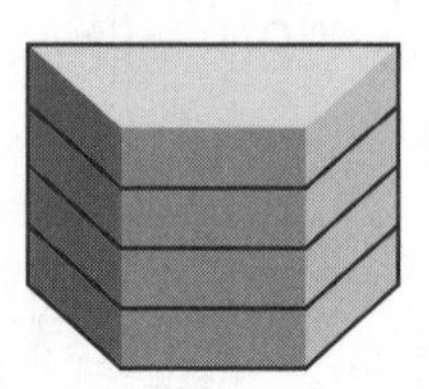
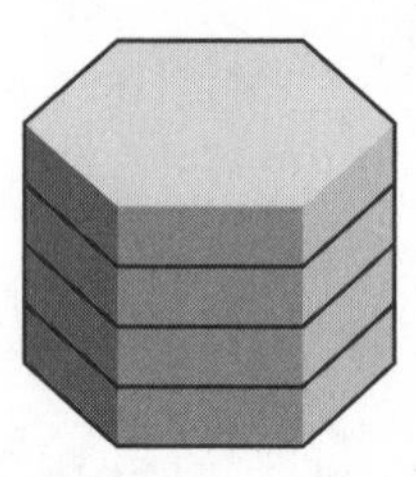

4. Below, carefully trace around each face of your prism. Then trace around each face of 2 or 3 more prisms on a separate sheet of paper. Share prisms with other people in your group. Ask someone in your group for help if you need it.

Math Boxes 6.12

1. Name each 3-dimensional shape.

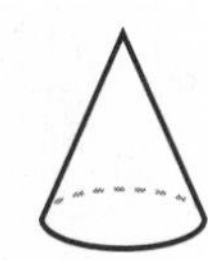

____

____

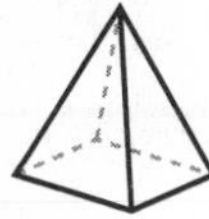

____

SRB 105 107 108

2. Draw a line, $\overleftrightarrow{AB}$, parallel to line segment $\overline{CD}$. Draw a ray, $\overrightarrow{EF}$, that intersects ray $\overrightarrow{GH}$.

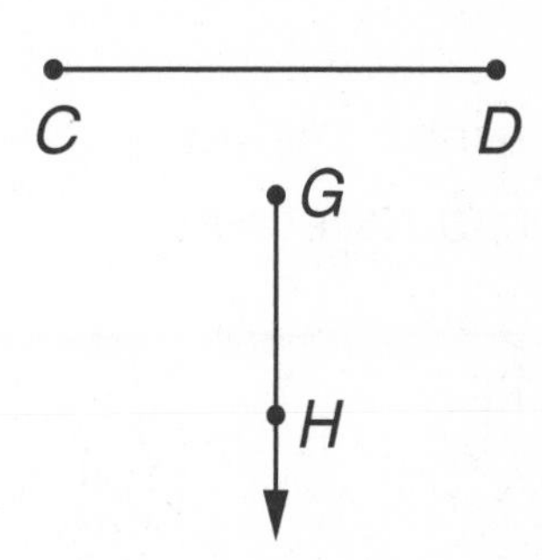

3. Give two reasons that this hexagon is not a regular hexagon.

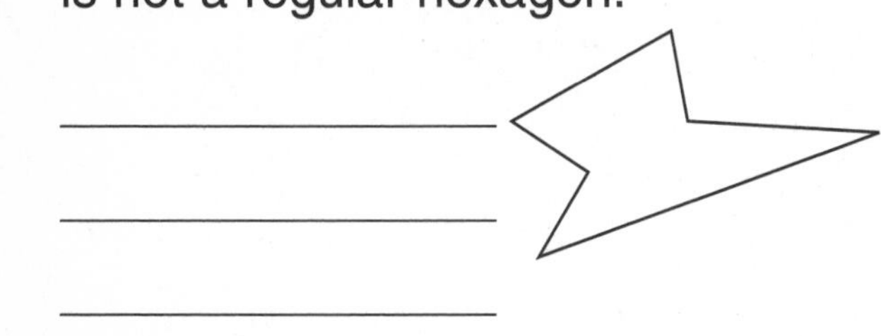

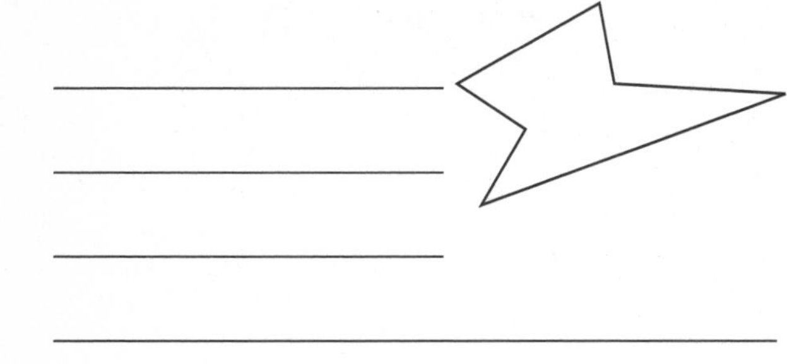

____

____

____

____

____

____

4. The degree measure of the angle is

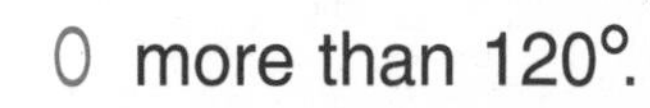

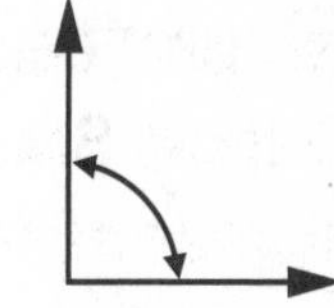

0 more than 120°.

0 less than 45°.

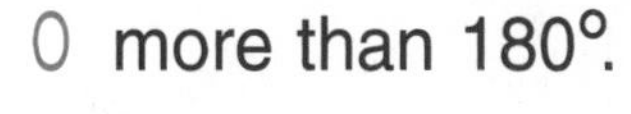

0 more than 180°.

0 90°.

5. What is a quadrangle?

____

____

____

____

____

____

6. Trace a figure from your template and draw the lines of symmetry.

The figure is a ______________.

It has _____ lines of symmetry.

Date Time

Math Boxes 6.13

1. Solve.

$2 \times 2 =$ _____

$5 \times 5 =$ _____

_____ $= 3 \times 3$

_____ $= 4 \times 4$

2. Circle the even numbers.

23,406 129

700,001 44,444

57 135,790

The numbers that are not circled are called __________ numbers.

3. Solve.

$5 \times 4 =$ _____

$2 \times 7 =$ _____

_____ $= 3 \times 10$

_____ $= 7 \times 10$

$3 \times 5 =$ _____

4. Continue the pattern.

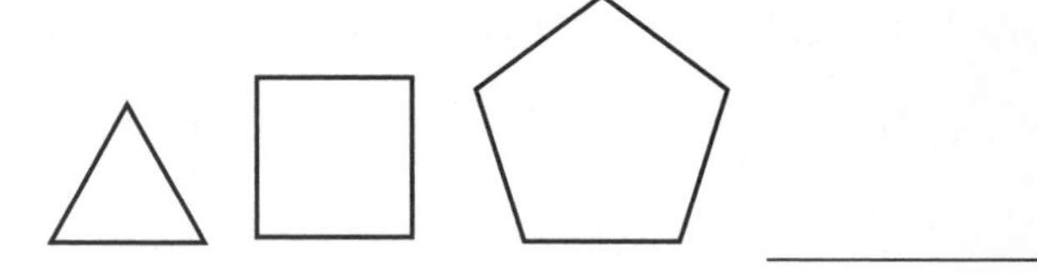

5. Complete.

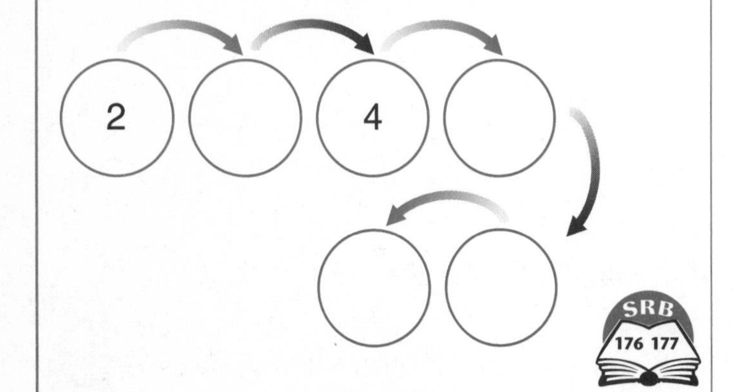

6. Write the fact family.

____

____

____

____

45
×, ÷
5 9

Date Time

Special Pages

The following pages will be used throughout the school year, first in this journal and then again in your *Math Journal 2* later during the year.

On the Sunrise and Sunset Record on journal page 158, you will record the date, and then the time of sunrise and the time of sunset for that date. You will begin to do this at the end of Unit 1 and then once a week or so whenever your teacher tells you.

Then later in the year, you will use the data that you have recorded on journal page 158 to make a graph on journal page 159. Your teacher will teach you how to do this in Unit 5.

Finally, on the National High/Low Temperature Project on journal page 160, you will record the following data: the U.S. city with the highest temperature and the U.S. city with the lowest temperature for the same date. You will do this every week or whenever your teacher tells you.

When you begin your *Math Journal 2* later in the school year, you will continue to record the sunrise and sunset times, and the highest and the lowest temperatures on pages in that journal. Near the end of the school year, you will use all this information.

Date Time

Sunrise and Sunset Record

Date	Time of Sunrise	Time of Sunset	Length of Day
			hr min
			hr min
			hr min
			hr min
			hr min
			hr min
			hr min
			hr min
			hr min
			hr min
			hr min
			hr min
			hr min
			hr min
			hr min
			hr min
			hr min
			hr min
			hr min
			hr min
			hr min

Date Time

Length of Day

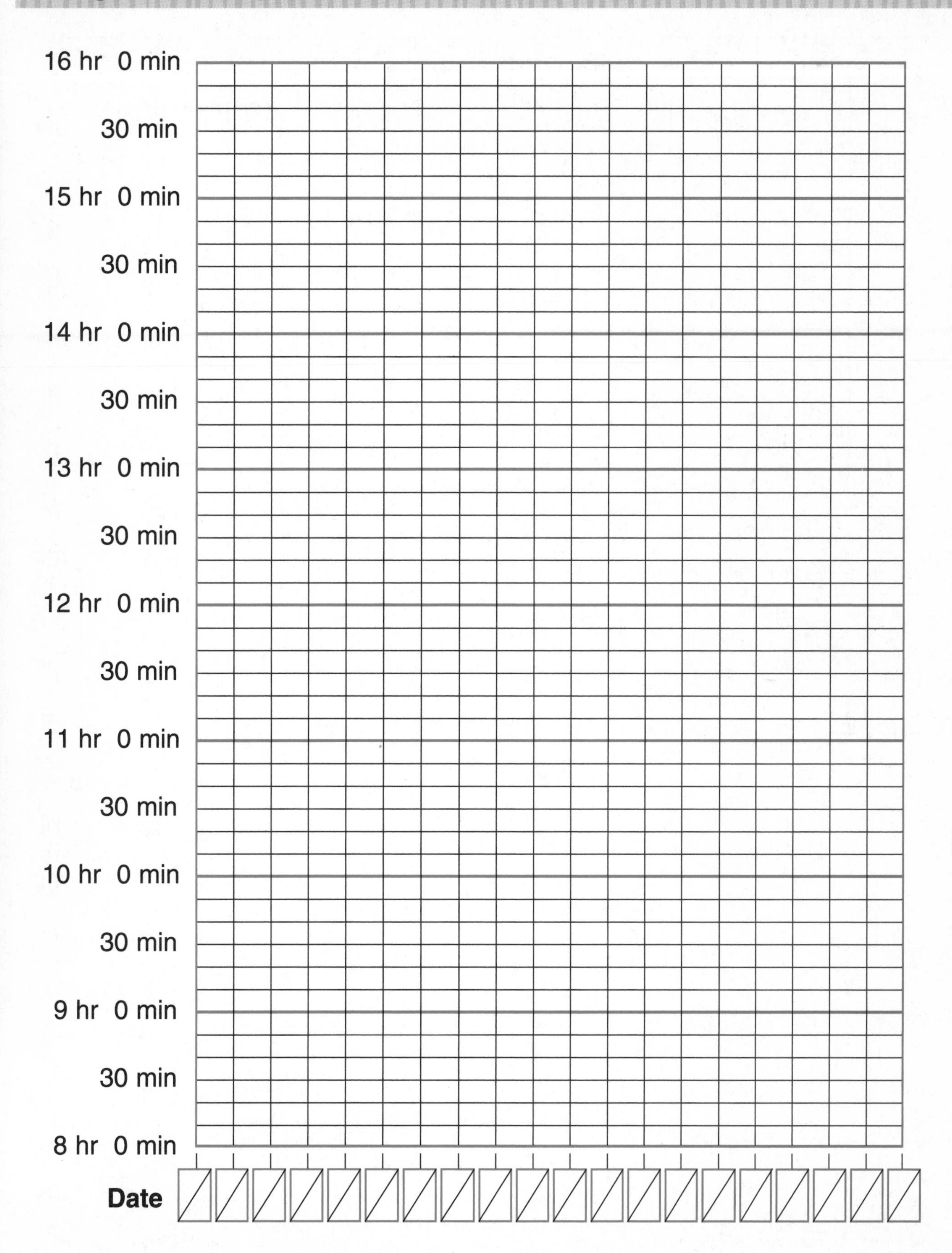

Date Time

National High/Low Temperatures Project

Date	Highest Temperature		Lowest Temperature		Difference in Temperature
	Place	Temperature	Place	Temperature	
		°F		°F	°F
		°F		°F	°F
		°F		°F	°F
		°F		°F	°F
		°F		°F	°F
		°F		°F	°F
		°F		°F	°F
		°F		°F	°F
		°F		°F	°F
		°F		°F	°F
		°F		°F	°F
		°F		°F	°F
		°F		°F	°F
		°F		°F	°F
		°F		°F	°F
		°F		°F	°F
		°F		°F	°F
		°F		°F	°F
		°F		°F	°F
		°F		°F	°F
		°F		°F	°F

Name Date Time

Paper Clock

1. Cut out the clock face, the minute hand, and the hour hand.
2. Punch a hole through the center of the clock face and through the Xs on the hands.
3. Fasten the hands to the clock face with a brad.

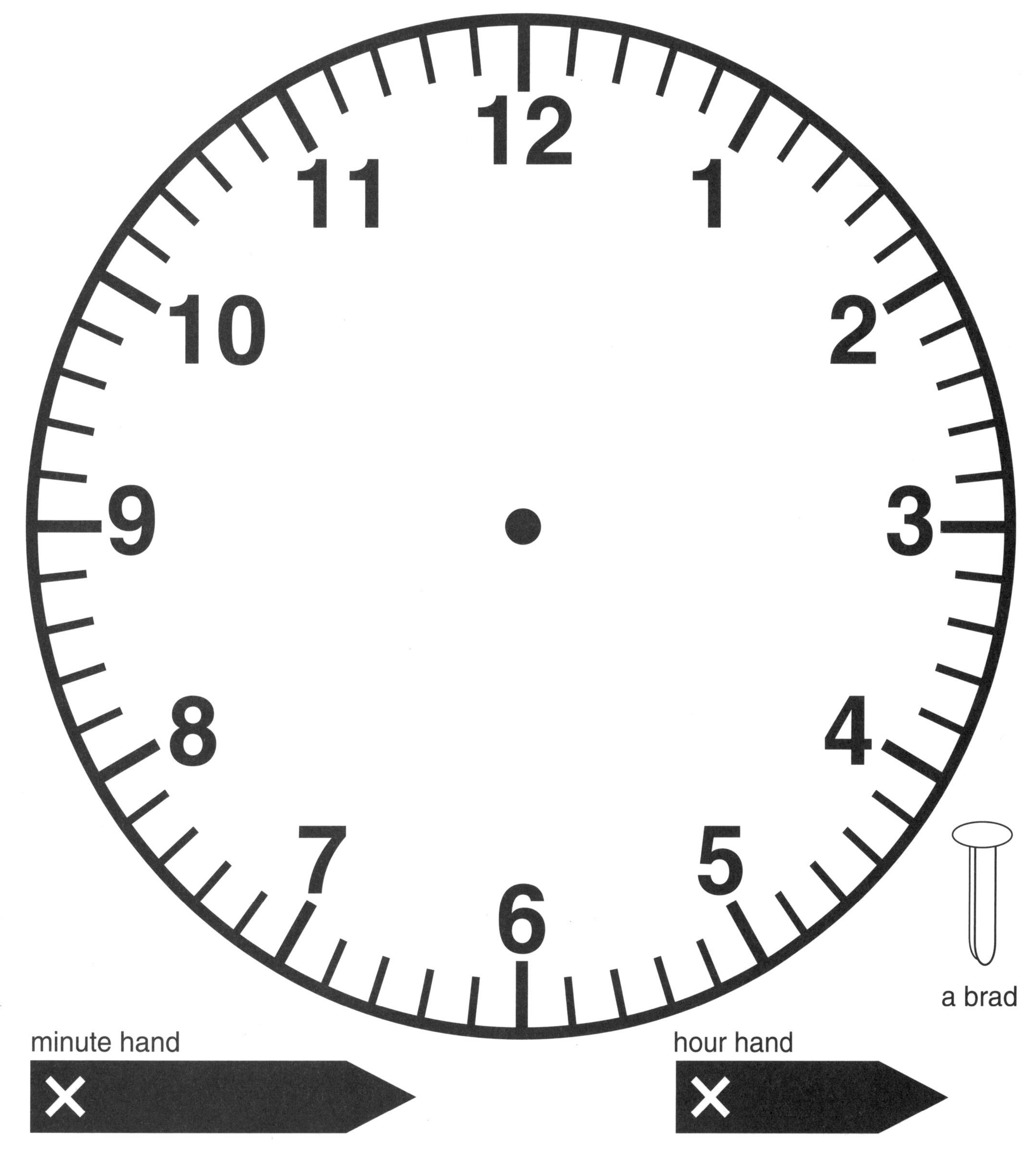

Name Date Time

Rulers

Cut out the rulers.

Inches (in.) 0 1 2 3 4 5 6 7 8 Ruler A

Inches (in.) 0 1 2 3 4 5 6 7 8 Ruler B

Inches (in.) 0 1 2 3 4 5 6 7 8 Ruler C

Inches (in.) 5 6 7 8 9 10 11 12 13 Ruler D

Centimeters (cm) 0 1 2 3 4 5 6 7 8 9 10 11 12 13 14 15 16 17 18 19 20 Ruler E

Name Date Time

Multiplication/Division Fact Triangles 1

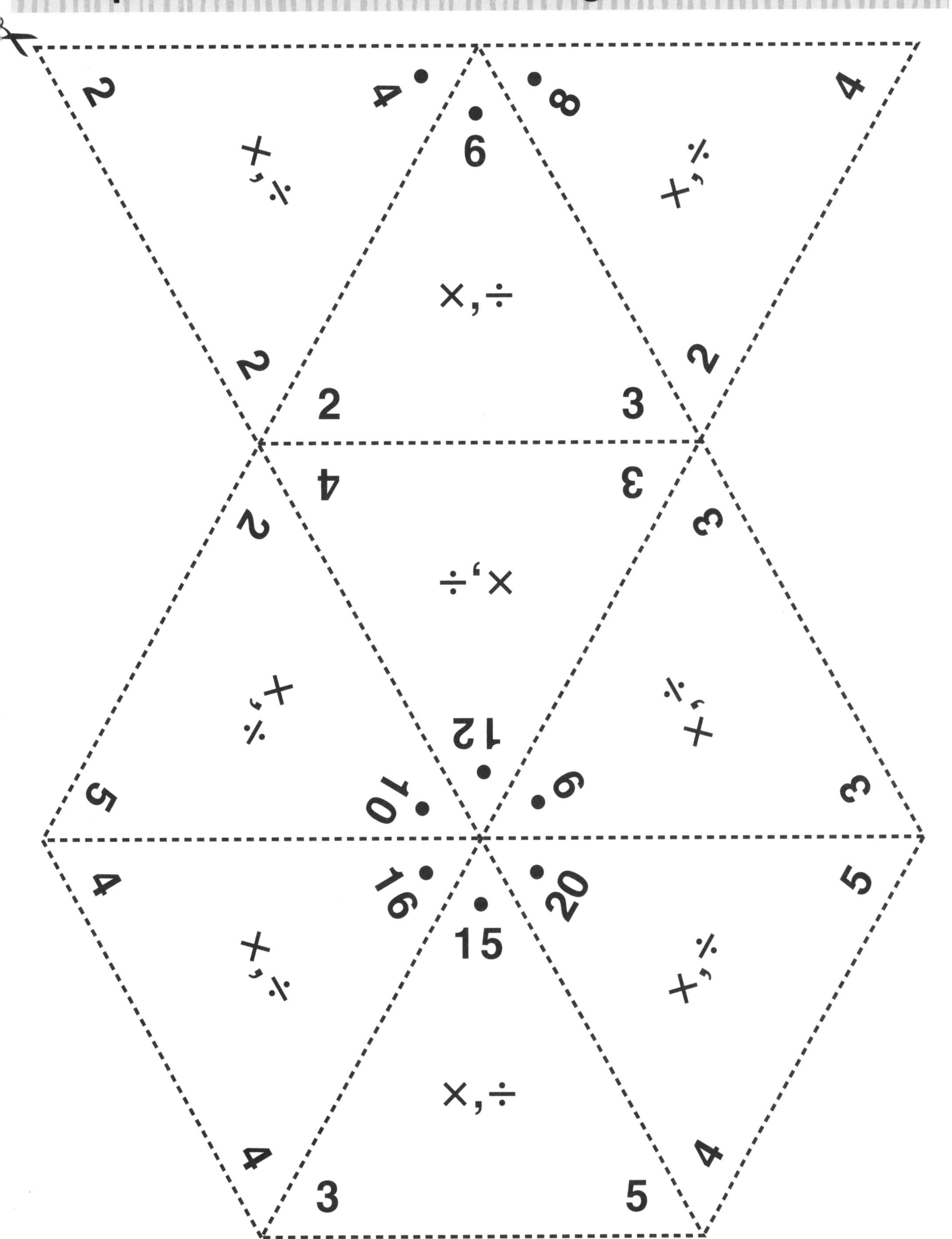

Use with Lesson 4.6.

Activity Sheet 3